Wastes Management Concepts for the Coastal Zone

REQUIREMENTS FOR RESEARCH AND INVESTIGATION

Prepared under the joint auspices of

Committee on Oceanography
NATIONAL RESEARCH COUNCIL
NATIONAL ACADEMY OF SCIENCES

Committee on Ocean Engineering
NATIONAL ACADEMY OF ENGINEERING

NATIONAL ACADEMY OF SCIENCES
NATIONAL ACADEMY OF ENGINEERING
Washington, D.C. 1970

ISBN 0-309-01855-2

Available from

Printing and Publishing Office
National Academy of Sciences–National Academy of Engineering
2101 Constitution Avenue, N.W.
Washington, D.C. 20418

Library of Congress Catalog Card Number 76-607568

PRINTED IN THE UNITED STATES OF AMERICA

Preface

Recognizing that a comprehensive evaluation should be made of the scientific and engineering requirements for research and investigation related to coastal wastes management, the National Academy of Sciences and the National Academy of Engineering, at the invitation of the Federal Water Quality Administration* of the United States Department of the Interior, initiated the study that resulted in this report. The subsequent development of this study was carried out under the auspices of the Committee on Oceanography (NASCO) and the Committee on Ocean Engineering (NAECOE), within the scope of their general concern for the further development and maintenance of our nation's understanding and use of the marine environment.

To plan the coastal wastes management study and prepare this report, the two Committees formed a Steering Committee co-chaired by Donald W. Pritchard of NASCO and Erman A. Pearson of NAECOE. At the initial meeting of the Steering Committee in January 1969, the goals of the study were established and a study session was outlined. Prior to the study session, 16 background papers in the primary areas of concern were prepared by invited experts as a basis for discussion and development of recommendations by the participants at the study session. These papers are available in a separate supplement to this report and are listed in Appendix D.

*Then, the Federal Water Pollution Control Administration.

The study session, held at Jackson Hole, Wyoming, on 7–12 July 1969, brought together a group of some 60 scientists and engineers from the nation's universities; municipal, county, state, regional, and national agencies; private consulting firms; and industrial research organizations. Their combined experience covered all aspects of wastes management research, planning, design, operation, and management. Some are involved directly or indirectly with a number of the nation's largest and most effective wastes management systems. Members of the Steering Committee, participants at the study session, and members of NASCO and NAECOE are listed in the Appendix.

Existing knowledge concerning management of wastes was reviewed, and recommendations were made for both basic research and design-related investigations that should be implemented to help ensure more effective wastes management in the coastal zone. Effective execution of an integrated wastes management program will require basic research and investigation in several areas, improved data collection and information processing, and the development of new instrumentation and sampling techniques. The recommendations contained in this report were developed and endorsed by the study session participants. These recommendations identify areas of need with respect both to research and investigation, and to wastes management practice; they should be considered as a basis for further detailed planning.

This report was prepared by the Steering Committee and was reviewed by the Committee on Oceanography and by the Committee on Ocean Engineering. It is not intended to serve as either a comprehensive treatise on, or a guidebook for, coastal wastes management practice. Sources of more detailed information have been identified in the Bibliography. The study session and the preparation of the report were made possible by the support of the Federal Water Quality Administration.

Our national activities and interests are becoming increasingly involved with the coastal zone and the deep ocean beyond. We hope that this report will be helpful in further developing the ability to protect and enhance the quality of the coastal zone and the marine environment.

Contents

Chapter 1
INTRODUCTION   1

Chapter 2
MONITORING   4
  Status of Information   5
  Waste Sources and Discharges   7
    Significant Discharges, 9; *Quantitation*, 10; *Methods*, 11;
    Waste Analyses, 15; *Selected Data*, 16
  Receiving Waters   19
    Parameters, 20; *Instrumentation*, 22
  Cost of Monitoring   22

Chapter 3
PHYSICAL PROCESSES AND INTERACTIONS   24
  Initial Dilution and Diffuser Design   25
  Physical Processes in Estuaries   26
    Circulation Patterns, 26; *Movement and Dispersion of an*
    Introduced Waste in Estuaries, 30; *Basic Relationships for*
    Dispersion in Estuaries, 31; *The Application of Models to the*
    Prediction of Dispersion of Wastes in Estuaries, 33
  Physical Processes in Coastal Areas   35
    Mean Motion, 36; *Dispersion*, 38; *Adequacy of Existing*
    Knowledge as a Base for Prediction, 40
  Floatables, Particulates, and Nonconservative Waste Components 41
    Floatables, 41; *Sinkables*, 41; *Nonconservative Waste Components*, 42

vi *Contents*

    Physical Effects of Waste Components on Receiving Waters   42
        Classification and Properties of Floatables, 43; *Formation of
        Slicks and Streaks*, 44; *Floatables*, 45
    Summary of State of Knowledge of Physical Processes   47
        Initial Dilution and Diffuser Design, 47; *Physical Processes in
        Estuaries*, 48; *Turbulent (Eddy) Flux*, 49; *Physical Processes in
        Coastal Areas*, 50; *Decay of Nonconservative Constituents as
        Related to Physical Factors*, 50; *Interactions between Floatable
        and Settleable Components of Wastes and Physical Factors*, 50

Chapter 4
CHEMICAL FACTORS   52
    Discussion of Recommended Fields of Research   55
        Inorganic Chemical Processes, 55; *Chemistry of Particles and of
        Processes in Sediments*, 56; *Nutrient Chemistry and Biochemical
        Changes*, 57; *The Chemistry of Specific Wastes*, 59; *Chemical
        Consequences of Man's Physical Activities*, 59
    Research Preserves   60

Chapter 5
BIOLOGICAL EFFECTS   61
    General Aspects   62
        Quantitative Analysis of the Biota, 62; *Assessment of Effects*, 63;
        Providing Guidance for Decisions, 67
    Specific Problems   68
        Pesticides, 68; *Public Health Risks*, 69; *Sludges and Solid Wastes*, 70;
        Temperature, 72; *Oil Spillage*, 74; *Toxic Substances*, 75;
        Nutrients, 77; *Dissolved Organic Compounds*, 79; *Oxygen Demand
        and Biological Degradation*, 79; *Brine Wastes*, 81; *Freshwater
        Discharges to Estuaries*, 81

Chapter 6
RECOMMENDED RESEARCH AND INVESTIGATION FOR EFFECTIVE COASTAL WASTES MANAGEMENT
    General Recommendations   84
        Concept and Criteria for Waste Treatment, 84; *Professional
        Development and Institutional Arrangements*, 84
    Recommendations Concerning Monitoring of Waste Discharges and
        Receiving Waters   85
        Research in Support of a Monitoring Program, 85; *A Monitoring
        Program*, 86; *Monitoring Waste Discharges*, 86; *Monitoring
        Receiving Water*, 88
    Recommendations Concerning Physical Processes and Interactions   90
        Initial Dilution and Diffuser Design, 90; *Physical Processes in
        Estuaries*, 91; *Turbulent Flux and Diffusion*, 92; *Physical
        Processes in Coastal Areas*, 92; *Decay of Nonconservative*

Contents vii

    Constituents as Related to Physical Factors, 93; *Interactions between Floatable and Settleable Components of Wastes and Physical Factors*, 93
    Recommendations Concerning Chemical Factors   94
        Chemical Processes Involving Dissolved Inorganic Constituents, 94; *Chemistry of Particles and Processes in Sediments*, 94; *Nutrient Chemistry and Biochemical Changes*, 95; *The Chemistry of Specific Wastes*, 96; *Chemical Consequences of Man's Physical Activities*, 97
    Recommendations Concerning Biological Effects   97

Chapter 7
SUGGESTED PRIORITIES AND ESTIMATED MINIMUM EFFORT REQUIRED   100
    Program Area of Monitoring Waste Discharges and Receiving Waters   101
    Program Area of Physical Processes and Interactions   101
    Program Area of Chemical Factors   101
    Program Area of Biological Effects   105

Appendix A   Steering Committee on Coastal Wastes Management   107
Appendix B   Participants in Coastal Wastes Management Study Session   108
Appendix C   Committee on Oceanography and Committee on Ocean Engineering   110
Appendix D   Background Papers   111

BIBLIOGRAPHY   113

Chapter 1

Introduction

The enhancement and maintenance of environmental quality has the attention of a worldwide public. Water quality, particularly that of estuaries and of coastal margin waters of the ocean, is of major concern. Here, at the interface between the developed land masses and the world ocean, intense development occurs to facilitate the use of the ocean as a major resource for food, minerals, power, transportation, recreation, and other benefits. This narrow environment must be maintained in a healthy and aesthetically attractive condition if the ocean is to continue to satisfy the multipurpose requirements of the human population.

Through this interface are transported most of the liquid wastes, and a growing fraction of the solid wastes, that result from man's activities. Although the ocean is vast, its ability to assimilate these wastes without significant degradation is not limitless. The waste products created by our increasing population are, in some cases, severely stressing the ecosystems in the waters over the continental shelf, which in their entirety comprise only 8 percent of the total volume of the ocean. Stress is also occurring in the other regions of the ocean.

As the quantity and variety of wastes being generated expand, global concern is being expressed about the contamination of the world ocean—especially by the oily materials of both shore and offshore origin, the so-called hard pesticides (of which DDT is the most notable example), the organomercury compounds, and lead. We need to expand the effectiveness of our existing wastes management systems, and to create new concepts and practices of wastes management.

Coastal waters, estuaries, and the open ocean have been the natural recipients of most of man's liquid-borne waste materials, as well as some of the atmospheric-borne and solid wastes. They will continue to be the ultimate recipient of the residual, nonreclaimable fraction of the wastes from man's activities.

Management of wastes is of course an element in the broader context of resource development and use. Coastal wastes management is effective if it implements policy and goals for resource use that encourage reuse of the reclaimable fraction of resources, and the appropriate treatment and introduction of the nonreclaimable waste fraction into the ocean environment. In some instances this can be done, not only without adverse effects, but even with a resulting beneficial effect on the environment, such as controlled fertilization.

Rather than simply encouraging greater disposal of wastes to the ocean, we must improve our understanding of the processes of interaction between the wastes and the environment and the characteristics of each. Significant and long-term deleterious effects can then be reduced substantially if adequate attention is given to wastes management systems that result in concentrations and time of contact that maintain and enhance the ecosystem.

Emphasis is required both on long-term research for understanding related oceanic processes, and on investigation of particular problem areas so that new wastes management concepts and facilities may be designed, operated, and evaluated.

A coordinated and comprehensive program to provide adequate long-term and short-term information for decisions on wastes management has two major aspects. The first aspect includes a research program to provide understanding and improve the scope, objective, and performance of waste discharge and receiving water-monitoring programs. Also included is an investigation consisting of the monitoring program itself. The other aspect comprises research and investigation concerned with the effects that physical, chemical, and biological processes have on the disposition of components of wastes introduced into the marine environment, and on the effects of the wastes on these processes in the receiving waters.

Those responsible and active in research, design, operation, and evaluation in wastes management are encouraged by the present level of public concern. However, effective expansion of the effort to enhance the quality of the coastal environment requires careful evaluation and determination of sequence and direction of effort, of expenditure rates, and of concepts and information resulting from research and investigation.

The program outlined in this report is the minimum effort recommended for improving the nation's ability to enhance the quality of the coastal waters and estuaries, and to continue the position of leadership it has attained in the development of information on, and techniques for, the effective management of wastes. This program of expanded effort will be effective only if it is conducted jointly by those in government—local, state, and national—and by those in industry who have responsibility for determining the goals and establishing the criteria for the management of wastes in the coastal zone.

Chapter 2
Monitoring

The effectiveness of waste management programs depends in major part on the scope, accuracy, and precision of the characterization of both the waste sources and the receiving waters. Rational waste control systems and facilities cannot be developed and operated without accurate information on the significant sources of waste, and on their relation to the receiving-water characteristics established for protecting the beneficial uses. Only in the light of this type of information can the limited financial resources available for waste control measures be effectively allocated.

Concerning the increasing concentrations of exotic wastes in the environment, the easiest and most reliable way of assessing the trend is by quantitative characterization of the waste streams where these wastes originate, or are presumed to originate, and where their concentrations generally are at their highest level. Problems associated with insensitivity of analytical methods and their accuracy and precision are usually reduced if not eliminated by this approach.

The foregoing comments about the need for accurate characterization of the waste streams, as well as of the receiving waters, appear logical and reasonable—at least to one familiar with water quality control technology. These concepts frequently become clouded, however, and in some cases completely obscured, in general public discussions of pollution. All too often this is also the case with uninformed scientists and engineers. The apparent reason is found in two general—and erroneous—concepts that can be oversimplified as follows:

1. That pollution-control workers are equipped with a work kit consisting of a large number of different sized "plugs"—their task being to search the environment for pipes discharging wastes and apply the correct size "plug" to each pipe.
2. That, with the existing technology of waste treatment, pollution control is simply the art of applying the "proper amount" of traditional treatment processes, primary, secondary, tertiary, etc., where the "proper amount" depends upon the discharger's financial ability, and one's belief in the "goodness" of traditional treatment methods.

Optimum progress and utilization of funds for wastes management systems require sufficient information for the effects on organisms of waste concentrations and contact time to be described quantitatively and for the concentration of various wastes to be attributed to their respective sources. Only then can rational technical judgments be made, with due consideration of the costs involved and of the necessity for reducing the input from particular waste sources to a specified rate. In some cases such reduction might mean their complete elimination.

Investigations of the effectiveness of existing and planned wastes management systems by the continuing characterization of waste discharges and of receiving waters are best implemented through monitoring programs. The effectiveness of the systems should be evaluated in the context of environmental quality. Particular attention should be given to maintaining the receiving-water quality characteristics associated with the particular location, and to enhancing and protecting the beneficial uses associated therewith.

Research is required to support and improve a monitoring program. Such investigation, related to the monitoring program, complements the recommended programs of research and investigation undertaken in physical, chemical, and biological processes.

STATUS OF INFORMATION

It is difficult to outline concisely the significant facets of a subject as diffuse as monitoring which includes evaluation of the sources and characteristics of wastes, measurement and recording of concentration levels of waste constituents in the receiving waters, and evaluation of environmental parameters. It is possible, however, to divide the overall problem into two parts:

1. Identifying the quantities and characteristics of the waste materials emitted by a particular type of waste generating activity.
2. Assessing the information available on the reactions and interactions of waste materials in the environment, and on the processes that alter their concentration profiles with time.

Much work remains to be done in relating quantitatively the effects of wastes to the concentration of the waste and the time of exposure (contact time) in the receiving water. Similarly, increased effort must be given to relating, empirically or otherwise, the waste concentration in the receiving waters to the mass emission rates of the various sources of waste discharges.

Table 1 lists the status of available information on waste sources and discharges, and indicates the areas where the information is generally adequate or inadequate. Similarly Table 2 lists the status of available information on reaction of wastes in the environment. These tables are not exhaustive, but they indicate generally the areas of needed investigation and research.

TABLE 1 General Status of Information on Waste Discharges for Waste Emission Rates and Characteristics

GENERALLY ADEQUATE INFORMATION

Flow rates of
    Municipal wastes
    Rivers and streams
    Storm–surface runoff
Common organic constituents—municipal wastes
    Biochemical Oxygen Demand (BOD) and Chemical Oxygen Demand (COD)
    Suspended and settleable solids
    Coliform bacteria
    Organic nitrogen compounds

GENERALLY INADEQUATE INFORMATION

New and/or exotic materials
    Toxicants—generally

TABLE 1 (Continued)

    Chlorinated hydrocarbons and biphenyls
    Lead, arsenic, and mercury compounds
    Petroleum and its degradation products
    Biostimulants including
        1. Conventional nutrients, N, P, and K
        2. Trace metals
        3. Organic growth factors, vitamins, etc.
    Taste- and odor-producing materials
Waste emission rates of natural runoff
    River discharges
    Surface (storm) runoff
    Agricultural drainage
Unit mass emission rates for significant wastes
    For cities (i.e., lb waste/person-day), etc.
    For industries (i.e., lb waste/10^3 lb product type), etc.
    For surface (and agricultural) runoff (i.e., lb waste/acre tributary area-day), etc.
Toxic and/or exotic wastes difficult to quantitate in environment
    Quantities of usage (materials balance) by drainage area
        1. Chlorinated hydrocarbons and biphenyls
        2. "Trace" toxic constituents (mercury, lead, etc.)
        3. Other trace organic constituents of concern
            Taste and odor materials
Technology for handling pulse releases and spills
    Near instantaneous discharge of batch process wastes
    Massive spills (pipelines, transport vehicles, etc.)
        1. Toxic materials
        2. Oils, petroleum products

WASTE SOURCES AND DISCHARGES

Scientists and engineers concerned with the general field of water quality control have acquired a broad base of knowledge and techniques applicable to the alleviation of certain waste problems facing society. However, appreciation of the magnitude of the total wastes management problem has been difficult for both members of the profession and the general public. Consequently, the tendency has been to concentrate most of the effort on "putting out fires" as they are recognized. A comprehensive, environmental quality concern and viewpoint has been slow in developing.

    The history of waste discharges into Santa Monica Bay on the southern California coastline illustrates this point:

TABLE 2 General Status of Information on Waste Discharges for Waste Reactions in the Environment

---

GENERALLY ADEQUATE INFORMATION

Organic, oxygen-consuming reactions (BOD, etc., carbon, nitrogen, sulfur cycles)

GENERALLY INADEQUATE INFORMATION

Decay rates of bacteria (coliforms, etc.)
Decay rates of exotic materials
    Pesticides, insecticides
    Pathogens
    Viruses
    Hydrocarbon fractions
    Toxic materials
Biostimulation factors (i.e., growth rate data)
    Conventional nutrients, N, P, K, etc.
    Trace metals
    Organic stimulants, growth factors, etc.
Toxicity
    Organics
    Heavy metals
    Pesticides

---

1. In early times, all raw wastes, though in relatively small quantities, were simply discharged into the nearshore waters of the bay in the belief that the discharge area was so far removed from the population that it constituted a "bottomless pit" where the problem could be forgotten.

2. As population and the quantity of waste discharges grew, and beach use increased, the area of biological hazard expanded and it became necessary to take measures to avoid endangering the beach users by exposure to water-borne diseases.

3. Continued population growth, and a better understanding of the overall waste problem, prompted treatment-plant design for both biological and aesthetic control.

4. The continuing increase in population, pressure from society for recreational use of the bay and its beaches, and concern for the ecology of the bay and coastline led to recognition of the fact that the total system needed intensive study.

5. In a basic investigation now being undertaken by the major waste dischargers in the region, the entire southern California nearshore water system is being studied as an ecological unit, including its interfaces with the atmosphere and land masses.

6. Ultimately, effects upon the world ocean must be considered.

Throughout this time it has been assumed that the ocean would remain the final wastes sink. In general, this is correct. Most conservative waste materials that reach waters at any place throughout the world will eventually be deposited in the oceans as the ultimate sink. This assumption does not necessarily imply disaster. For example, in Santa Monica Bay, even with the volume of liquid wastes now being received, the water quality levels that were established by the State of California Water Quality Control Board for all of the uses of the Bay, including water contact sports and fishing, are being maintained.

The problems of waste discharge into coastal waters from other, relatively uncontrolled sources such as agricultural runoff, and of the rapidly increasing waste loads arising from the increase in population make a constant review of treatment requirements and of coastal water-quality criteria and assimilation capacity necessary. It is not sufficient simply to stand on past practices; the environmental quality problem is not static, and the level of research, investigations, and monitoring must reflect its increasing complexity.

Significant Discharges

Specifications for a core, minimum monitoring program require the definition of "significant" waste discharges, to which the program is to be applied. A research program is necessary to develop a consistent and usable definition of "significant." In this discussion, significant waste discharges are considered to include, but not necessarily be limited to, the following candidate waste materials:

1. Municipal and industrial waste streams.
2. Storm runoff and combined sewer overflows.
3. Water courses containing significant waste materials.
4. Batch waste dumping and barging operations.

It should be recognized at the outset that all waste discharges—especially minor ones such as the treated, strictly domestic wastes from 100 persons discharging into open coastal waters—are not classified as "significant." On the other hand, many major waste discharges require much more analysis than the core minimum program for proper characterization.

Considerations involved in determining whether a discharge is significant include:

1. The magnitude (flow and waste-mass emission rates) of the discharge as compared with:

    a. The available dilution and quality requirements of the receiving waters;

    b. The relative magnitude of the discharge compared with other discharges in the general area; and

    c. The defined or undefined character of the effect of the waste on the receiving water and beneficial uses.

2. The relative cost of conducting the minimum "core" characterization program as compared with:

    a. The cost of at least secondary treatment for the waste discharge;

    b. The cost of alternate methods of disposal; and

    c. The potential damage by the discharge.

Specifications for the "significant discharge" category must be sufficiently general so as not to exclude some specific and significant discharges of potential ecological damage.

Quantitation

Quantitative data are needed on both municipal and industrial waste discharge characteristics. Waste streams should be metered, sampled, and analyzed to yield statistical descriptions of the significant waste mass emission rates (MER) (e.g., lb grease/day). The waste mass emission rates provide the basis for the unit mass emission rate (UMER) which is the most important waste discharge characteristic that can be interpreted and utilized in waste management decisions. The UMER is defined as the MER per unit of waste-generating activity.

The UMER can be readily determined if accurate information is obtained on the level of waste-generating activity. For municipal waste streams the waste-generating activity unit is the individual; that is, the UMER is the MER per tributary (connected) population. With industrial wastes, the waste-generating activity is some suitable unit of production (e.g., 10^3 lb of specific product). The product should be identified; and it should be the direct result of the waste-generating activity. A typical form of expression for UMER for both municipal and industrial wastes is shown below:

UMER (phenol)   Primary Effluent   Settled Refinery Waste
               1×10^{-4} lb/capita-day 5×10^{-3} lb/10^3 lb gasoline

UMER data are of major value for all waste-discharge descriptions and especially for industrial wastes. They permit the following analyses and comparisons that otherwise could not be made.

1. It is possible, using UMER, to compare the effective performance of various types of municipal waste-treatment systems and to make reasonably accurate benefit-cost analyses.

2. The magnitude and variation in the UMER provide the best practical estimate (short of an exhaustive and extremely expensive industrial waste survey of the municipal sewer system) of the relative contribution of industrial wastes in a combined, municipal-industrial sewer system (the usual metropolitan sewerage situation).

3. The magnitude and variation of the UMER and MER provide a meaningful check on the extent and uniformity of operational control of waste treatment systems.

4. For discrete, industrial waste discharges, the UMER provides the only meaningful information available on the level or degree of waste elimination and treatment within the waste-generating activities.

5. Data on industrial UMER for comparable or similar waste-producing processes in different plants provide a basis for comparing the relative waste-handling practices in the different plants. It should provide a rational basis for benefit-cost analyses of waste control measures.

6. Statistically valid data on UMER from an industrial plant with several waste-producing processes permit identification and evaluation of the principal process sources of significant waste emissions. Multiple regression analyses, made with the aid of digital computers, are employed. The alternative would be to make a MER analysis of each process waste stream—an expensive operation.

Methods

Adequate characterization of waste streams requires that attention be given to three independent but important aspects of the problem.

To develop reliable data on waste mass emission rates (MER) and unit emission rates (UMER), one must have (a) accurate data on the volume flow rates (F), (b) representative samples of the waste stream, and (c) accurate analytical methods for determining the pollutant concentrations in the collected waste sample. The objective, generally, is to determine the mean waste mass emission rate. This quantity is the product of the mean flow rate (F) times the mean concentration of the waste in the waste stream. Determining the mean waste concentration in waste streams requires the collection of representative samples for analysis. This can be done by withdrawing continuously a sample from the waste stream (or collection of an aliquot) in direct proportion to the flow rate during the interval of sampling.

The general problem of composite sampling in proportion to flow has not been given adequate attention in most treatment plants, particularly the smaller ones. However, even many of the larger treatment installations use only "grab" samples for analysis. Because of the manner of sampling, as well as the retention time through process for a variable-concentration feed stream, the results of such "grab" samples may be as much as 30 to 100 percent in error.

An additional complicating factor is the fact that waste streams are rarely, if ever, at steady state; the flow and the waste concentrations are quite variable, especially with industrial wastes which frequently result from batch-type, process discharge rather than from continuous flow processes. Consequently, the problem of sampling proportional to flow to obtain mean concentrations of the contaminant in the samples collected needs serious consideration.

The domestic waste stream is not as uniform as some think, and its variation must be considered if one is to obtain reliable estimates of MER. The normal daily variation in domestic sewage flow rates is from one half to double the average flow. In addition, there is an organic waste concentration variation from half to double the average level. Surprisingly, the minimum waste concentration occurs at the same time that minimum flow occurs, and the maximum waste concentration usually occurs during the daily maximum flow period. Thus, the MER data obtained from grab samples could vary by a factor of four depending upon the time of sampling and flow measurement.

METERING

Collection of samples for analysis on a composite basis requires the use of accurate and reliable flow meters. Those with mechanical rotors, such as propellor meters, often offer substantial maintenance problems. Venturi meters, venturi flumes (PB meters), and magnetic meters are most suitable. However, venturi meters should be equipped with clean-out taps for the pressure-transmitting lines. In most situations the meter should be connected to a continuous recorder and totalizer or flow integrator. Integrators that can be set to activate an external circuit at any predetermined increment of flow integration—500 gal, 5,000 gal, and 20,000 gal—are the preferred recorder integration combinations. As is usual for flow meters, an accuracy of measurement of ±3 percent is preferred with ±5 percent the outside limit.

SAMPLING

Among the most neglected phases of waste monitoring have been the type, mechanics, accuracy, and precision of waste-stream sampling methods. The representative sampling of waste streams is not a simple task. Waste constituents of concern include not only the substances in solution but also the suspended-solids fraction consisting of both fine and coarse material. The floatable matter consists of oil and grease as well as buoyant particulates. The physical and chemical properties of wastes such as oils and grease may make it difficult to obtain aliquots in which the concentration of wastes in the collected sample is representative of the flow-weighted concentration in the waste stream.

It has not been possible to locate a significant study, either theoretical or practical, that has considered or described the errors associated with variable-pulse or interval-type sampling systems for waste streams that have significant, defined transients in flow rate or pollutant concentrations alone or in combination. Such a study is needed to provide specifications or a guide for proportional waste-sampling systems.

COMBINED SYSTEMS

In recent years many near-continuous, proportional-type sampling systems have been available to supplement the early dipper-type sampler used for open-channel flow conditions. Most of these samplers employ variable speed pumps or dippers, geared in proportion to waste flow rates. The samplers are often followed by some type of flow-splitting device to reduce the sample volume to manageable proportions. In most of them there are limitations on sample flow rates that affect the representative character of the selected sample.

In the early 1950's one of the most economical, dependable, and representative proportional-flow sampling systems known to date was developed, but it has not been widely used. The details of this sampling system are shown in Figure 1. The proportional sampling station consists of readily available components—a venturi flow meter and recorder-integrator connected to a solenoid-activated sample line deflector; a sample bypass line flowing continuously, with a short flexible section; a sample receiver; and refrigerated storage containers. The flexible section of the sampling bypass line, normally in the bypass position, is

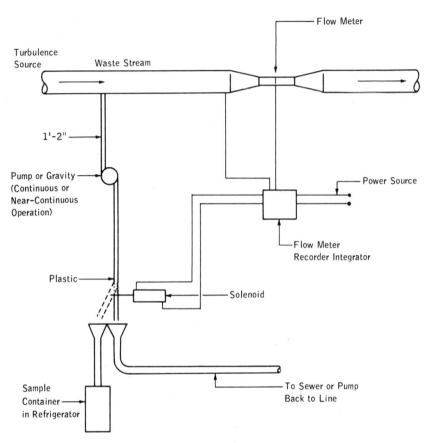

FIGURE 1 Proportional composite process stream sampling system.

deflected for a few seconds (adjustable) by a solenoid-activated piston which is triggered by an electrical pulse from the flow recorder-integrator. During the few seconds deflection, a constant sample volume is directed to the sample container. The flow integrator can be set to activate the solenoid for any preselected increment of waste volume passing the flow meter. This volume will depend upon both the flow variation in the waste stream and the waste concentration variation. With this apparatus, it is possible to collect a representative sample for any quantity of flow passing the metering station.

Experience with many installations of the type described indicates that it is possible to select a combination of period of sample withdrawal and unit of flow integration that will yield a representative, proportional sample of the waste stream.

Waste Analyses

One of the vexing problems associated with a waste monitoring program is selection of the analyses to be included. The parameters to be monitored depend upon the general character of the waste stream, the type of waste treatment, and the character and desired use of the receiving water. The exact mix of analyses to be used is related to the local circumstances. There should, however, be a near-minimum set of analyses that ought to be conducted on all significant waste streams. The final selection of parameters, or analyses, to be included in any monitoring program will be a compromise between what is needed, and what can be supported by available funds.

Selection of the parameters to be considered in a waste inventory program should be guided by the following principles:

1. The central core of parameters should be those that are considered absolutely essential in either the waste treatment, disposal planning, and design function, or are of known or presumed practical significance for interpreting or evaluating the effects of the wastes on the receiving waters.

2. Analyses of second-level significance can be included, considering both the cost of the analyses and the ability to interpret the results.

3. Serious consideration must be given to the addition of newer or less conventional waste parameters at the expense of dropping some of the classic or traditional, but seemingly uninterpretable, parameters—at least insofar as the design or evaluation function is concerned. Specific examples of some analyses that might be added as well as those that might be deleted from routine waste monitoring programs are:

Possible Additions	*Possible Deletions*
Toxicity (48 or 96 hr median tolerance limit)	Total Solids
	Complete Mineral Analyses[†]
Biostimulosity *	Alkalinity
Floatable Matter	Turbidity
Pesticides (Hard, i.e., Half Life $>$ 6 months)	Settleable Solids
Threshold Odor Number	

*The concentration required to increase selected algal growth rate in chemostat by 50 percent or some similar growth rate characteristic.
[†]Most ions such as Ca, Mg, K, Na, SO_4, plus SiO_2.

4. Consideration should be given to reducing the number of analyses that measure essentially the same thing, and to eliminating those that studies have shown are highly correlated with other selected analyses for a particular waste stream.

5. For every parameter or analysis included in the monitoring program, information should be available or developed on the accuracy and precision of the sampling method as well as the laboratory analyses.

Monitoring programs applied to waste discharges must be reviewed frequently to make certain that significant analyses with respect to effects and concentrations of concern in the receiving waters are included. In practice, unfortunately, many monitoring programs simply have added new parameters for analysis without the critical review needed to delete inconsequential or redundant analyses.

Selected Data

To illustrate both the general magnitude of the problem and the type of data that have been collected for typical and major waste disposal operations, several examples of mass emission rate data are presented in Tables 3 through 6. Table 3 presents the combined, average, liquid-borne, waste-mass, emission-rate data for the discharges of the Los Angeles County Sanitation Districts (Whites Point—360 mgd flow) and the City of Los Angeles (Hyperion Treatment Plant—340 mgd) for the year 1968. These two installations serve a total tributary population of about 7 million persons.

It is of interest to compare the magnitude of waste constituents discharged to the Pacific Ocean with those discharged into the atmosphere of the Los Angeles basin. Table 4 is a summary of available waste mass emission rates to the atmosphere as reported by the Air Pollution Control District of Los Angeles County. In addition to the liquid-borne and the atmospheric wastes, the same population produces 30,000 dry tons/day of solid wastes (e.g., trash).

Note that in the Los Angeles County area more organic wastes are discharged into the atmosphere than into the Pacific Ocean. Obviously, the types of organic matter differ considerably in chemical character. Possibly of even greater interest are the reported data on nitrogen discharges. Nitrogen oxides discharged to the atmosphere on a nitrogen basis amount to over twice the reported nitrogen discharged directly to the Pacific Ocean.

Tables 5 and 6 report selected unit mass emission rates for municipal waste discharges from different types of treatment processes as reported

TABLE 3 Combined 1968 Discharge of the Los Angeles County Sanitation Districts (360 mgd) and City of Los Angeles (340 mgd) to the Pacific Ocean

Constituent	Mass Emission Rate (tons/day)	Unit Mass Emission Rate (lb/capita-day)
Dissolved solids	3,600	1.03
Chloride (Cl)[a]	1,150	0.330
Sodium (Na)[a]	880	0.252
Sulfate (SO_4)	610	0.174
Suspended solids (includes digested sludge)	565	0.162
BOD	560	0.160
Total nitrogen (N)	165	0.047
Phosphate (PO_4)	100	0.029
Grease	90	0.026
Potassium (K)	82	0.023
Thiosulfate (S)	50	0.014
Detergents	21	0.006
Phenols	9	2.57×10^{-3}
Iron (Fe)	7	2.00×10^{-3}
Fluoride (F)	5	1.43×10^{-3}
Boron (B)	4.4	1.26×10^{-3}
Zinc (Zn)	2.4	0.69×10^{-3}
Chromium (total Cr)	1.3	0.37×10^{-3}
Copper (Cu)	1.0	0.29×10^{-3}
Selected Trace Constituents	(lb/day)	
Nickel (Ni)	1,125	1.60×10^{-6}
Cyanide (CN)	930	1.32×10^{-6}
Lead (Pb)	730	1.04×10^{-6}
Manganese (Mn)	400	0.57×10^{-6}
Cadmium (Cd)	270	0.38×10^{-6}
Arsenic (As)	100	0.14×10^{-6}
Selenium (Se)[b]	55	0.16×10^{-6}
Chromium (hexavalent, Cr^{+6})	30	0.043×10^{-6}
Barium (Ba)[b]	∼0	∼0
Silver (Ag)[b]	∼0	∼0

[a]Does not include direct discharge of oilfield brines to ocean.
[b]Data for City of Los Angeles only.

by the University of California's Comphrehensive Investigation of San Francisco Bay. These tables, which report unit mass emission rate data for Biochemical Oxygen Demand (BOD_5) and Chemical Oxygen Demand (COD), are representative of the waste discharge emissions for some 20 waste parameters outlined in the report of the Comprehen-

TABLE 4 Waste Mass Emission Rates of Los Angeles County Basin to Atmosphere[a]

Constituent	Mass Emission Rate (tons/day)
Organic gases, mainly hydrocarbons	2,500
Particulates	110
Nitrogen oxides	950
Sulfur dioxide	225
Carbon monoxide	9,695
Total (all sources)	∼14,000

[a]Data from Air Pollution Control District of Los Angeles County.

TABLE 5 Unit BOD_5 Mass Emission Rates[a] for Municipal Treatment Plant Discharges

Type of Treatment	Type of Waste (lb/capita-day)		
	Principally Domestic Waste	Domestic plus 20% Cannery Waste	Domestic plus 20% Industrial Waste
Primary	0.0838[b] (0.131)	0.102 (0.185)	0.120 (0.200)
Single-stage trickling filter	0.0283 (0.0340)	0.102 (0.105)	0.744 (0.0660)
Multistage trickling filter	0.0117 (0.0137)		0.0144
Trickling filter activated sludge	0.00827 (0.00860)	0.121 (0.128)	
Activated sludge	0.0506 (0.0393)		

[a]All values are geometric mean values.
[b]First number is geometric mean of all data in group. Number within parentheses is annual mean of corrected data adjusted to remove the effect of time of sampling and possible bias due to varying number of samples collected during the year.

TABLE 6 Unit Chemical Oxygen Demand Mass Emission Rates[a] for Municipal Treatment Plant Discharges

	Type of Waste (lb/capita-day)		
Type of Treatment	Principally Domestic Waste	Domestic plus 20% Cannery Waste	Domestic plus 20% Industrial Waste
Primary	0.235[b] (0.340)	0.291 (0.480)	0.322 (0.560)
Single-stage trickling filter	0.121 (0.107)	0.253 (0.265)	0.238
Multistage trickling filter	0.0874 (0.0880)		0.0976 (0.0970)
Trickling filter activated sludge	0.0546 (0.00860)	0.386 (0.400)	
Activated sludge	0.174 (0.155)		

[a] All values are geometric mean values.
[b] First number is geometric mean of all data in group. Number within parentheses is annual mean of corrected data adjusted to remove the effect of time of sampling and possible bias due to varying number of samples collected during the year.

sive Investigation. Similar types of data are reported for the significant industrial discharges in the San Francisco Bay area.

RECEIVING WATERS

Monitoring the coastal marine environment for waste components and their effects on this environment should be considered on the basis of a total system concept. In this regard, a monitoring system should serve the following functions:

1. Provide intermittent or continuous characterization of a limited body of water and its terrestrial and atmospheric interfaces. This goal may be accomplished by means of pertinent physical, chemical, or biological measurements sufficient to define the significant nature of the water body throughout a limited time period specified on the basis of statistical validity.

2. Provide information on the significant sources of mass movement and their residence time in the limited water body, establish the signifi-

cant character of such sources, and evaluate the relative contribution of each to the nature of the water body.

3. Provide for rapid data evaluation and indicate the response procedures appropriate for the given water condition.

Monitoring, therefore, should include consideration of data collection, evaluation, and response procedures with a view toward reduction of the elapsed time interval between the first and last of these actions.

It is recognized that monitoring the coastal environment is difficult. Moreover, the estuarine region—the intermediate zone between freshwater rivers and open ocean—is affected by the mass movements of each, but possesses the character of neither. Consequently, monitoring the estuarine region is even more complex.

The proximity of estuaries to population centers and their unique characteristics as natural resources, recreational assets, and habitat and conditioning areas for marine life demand their protection from pollution. Providing such protection requires the ability both to define and measure pollution parameters and to predict and prevent the adverse effects of waste discharge.

Consequently, efforts to determine the criteria necessary for the accurate characterization of a water body must also consider the availability or state of development of rapid, accurate procedures for measuring the parameters selected. In addition, instrumentation may need to be developed specifically to perform the selected analyses and to transmit or record the data. Finally, data analysis techniques (preferably computer methods) for rapid data evaluation should be utilized to provide the basis for timely response or control procedures. The final control or corrective procedure should be the ultimate output of the monitoring program. However, the control action is usually separated from monitoring and data evaluation procedures.

The evaluation of waste discharge effects usually requires multivariate analysis techniques. Wastes may significantly change the character of the receiving water. Monitoring is generally based upon either determination of the degree of change in particular characteristics of a body of water, or on the concentration level therein of waste constituents that are known or presumed to be responsible for significant changes in these characteristics.

Parameters

Among the fundamental problems that must be solved by a monitoring program are (a) detection of a specific response in receiving-water quality characteristics caused by a particular waste mass emission rate, and (b) assessment of the statistical level of significance of the response. Thus, the parameters to be monitored not only should be those that reflect important and critical environmental characteristics of the water body, they should also be the ones most directly affected by variations in the waste load that is imposed.

Historically, in the marine environment the most frequently used water quality parameter has been the coliform test for the detection of the relative concentration of domestic sewage that is present and the assessment of the apparent health risk. It is now recognized, however, that the coliform concentration has become more of a criterion of aesthetic acceptability and attainability than an accurate measure of health risk. In routine monitoring, in addition to coliform organisms, several general environmental parameters such as temperature and transparency and, in estuarine systems, chlorosity and dissolved oxygen have been included.

In recent years, along with the increasing concern for environmental quality, attention has been focused upon criteria that better describe the aesthetic and biological condition of the marine environment. This emphasis has resulted in the occasional inclusion of aesthetic and biological parameters in monitoring requirements. However, the full application of such parameters is in its infancy. For example, there is no current, quantitative, analytical method for assessing the amount of floatable material (including oil) that is present; thus, analytical comparisons of receiving-water conditions with emission rates for floatable material are not possible. Yet, the aesthetic appearance and condition of receiving waters is one of the public's greatest concerns.

Assessment of the biological condition, or health, of the receiving waters is one of the most critical, complex, and controversial areas in the waste management field, especially with respect to what can and should be done with monitoring programs. There is a critical need for accurate and simple methods of assessing the general condition of the biota, including benthos, plankton, and, if possible, the significant fisheries.

It is beyond the scope of this discussion to address the general subject of biological indexes and parameters. However, the design of a biological monitoring program should consider the quantitative assessment of enrichment as well as of toxification. Emphasis should be given to sedentary organisms that are easier to collect than motile forms and that provide an integrated response to water quality factors over a considerable time period. Considerable research should be conducted to determine the most sensitive and practical assaying methods. Consideration must be given to the use of the benthic fauna as well as the plankton for this purpose. Information will have to be obtained on the physical and chemical characteristics of the sediments, as well as of the overlying water, when the benthic fauna are used.

From a review of the literature and recent experience with monitoring studies, it appears that the following parameters and indexes of biological effects and of health deserve further study and field verification:

1. Bioassays, mortality, and physiological response rate.
2. Diversity indexes.
3. Similarity and dissimilarity coefficients.
4. Productivity/respiration.
5. Biotic indexes.

Instrumentation

To make monitoring programs most effective and economical, considerable effort will have to be given to the development of suitable instruments for measuring, transmitting, and recording water quality data. Meters and recorders for physical parameters such as water and wind currents, temperature, and conductivity are available, although minor modifications may be necessary for certain applications.

The state of development of instruments for continuously recording chemical parameters is in its infancy, even for such common substances as organic carbon. Considerable instrument development work will be required for any of the meaningful chemical parameters where the character of the problem justifies such equipment. As for biological monitoring on a continuous or near-continuous basis, this subject has not yet received any significant attention, primarily because of the lack of a consensus as to what parameters should be measured and what they mean.

COST OF MONITORING

In a discussion of the monitoring of waste discharges and receiving waters, some consideration must be given to the cost of such programs. Obviously, there can be no standard formulation as to just how much effort should be devoted to monitoring. In most instances, however, the subject of monitoring to assess performance of processes and of treatment plants has been grossly neglected.

When adequate methods are available for monitoring the meaningful parameters of waste discharges and receiving water, the responsibility for the financing and the conduct of the monitoring program must rest primarily with the waste discharger. Effective waste and environmental monitoring is a justifiable, although minor, part of the cost of waste disposal.

One reasonable basis for estimating the amount of effort that should be devoted to environmental monitoring would be a fraction of the cost of the waste-treatment processes that are justified for protection of the environment. Certainly, a few percent of the annual waste-treatment costs should be devoted to assessing performance of the plant and the impact of the wastes on the receiving waters. Considering that the unit cost of waste treatment may range from as low as $40 per million gallons of waste treated (large plants, minimum treatment) to as high as $300–400 per million gallons (small plants, high degree of treatment), 2 percent of such expenditures would range from $0.80 to $8 per million gallons. This cost would amount to some $290/day for receiving-water monitoring at a plant the size of the one operated by the Los Angeles County Sanitation District, or about $50/day at the Ocean City, Maryland, plant.

It may be of interest that the routine practice of receiving-water monitoring in California has reached a point such that a 1964 report on the evaluation of scientific and technical data on marine waste disposal for the State Water Quality Control Board included an empirical correlation between the annual cost of receiving-water monitoring and the size of the discharge in millions of gallons per day (mgd).

Soundly conceived monitoring programs can yield much information that will assist in guiding the direction of research. The funds for the research required to improve analytical methods for the monitoring programs should not be directly related to the funding of new waste-treatment systems and facilities.

Chapter 3
Physical Processes and Interactions

Physical processes in estuaries and coastal waters are major factors in the distribution of wastes in these waters. In turn, the waste materials may alter the physical properties of the receiving waters.

Consider, for example, the fate of a conservative waste introduced in solution or colloidal form into estuarine or coastal waters. Its subsequent distribution over a time period will depend on the manner in which the waste is initially introduced and on the initial difference in density between the undiluted waste solution and the receiving water.

If the waste is introduced as a high velocity jet, momentum entrainment will mix the volume containing the wastes with the receiving waters, rapidly diluting the wastes. If, as is usually the case in estuarine and marine coastal waters, the waste stream is less dense than the receiving water, the waste volume will be buoyant and, if introduced at the bottom, will ascend toward the surface, entraining additional diluting water en route. If the wastes are discharged at sufficient depth, in waters having a vertical density stratification, the ascending waste volume may not reach the surface, but instead will be diluted to the point of neutral buoyancy at some intermediate depth, where its vertical movement will cease. If the waste volume is initially denser than the receiving waters, and is introduced at the surface, it will sink, entraining dilution water en route to the bottom or to some intermediate depth in a manner similar to the ascending, buoyant volume.

Following initial mechanical mixing, the somewhat enlarged and diluted waste volume will be subjected to the complex physical processes of movement and dispersion. The term "dispersion" refers to a multitude of physical processes occurring internally in a body of water which tend to produce uniformity in the spatial distribution of properties, or the spreading of a patch of waste material.

Two processes contribute substantially to dispersion: "advection" and "diffusion." In advective processes, large-scale, regular patterns of water movement, such as the net estuarine circulation, carry wastes with them, thus producing a local change of concentration with time. In diffusive processes, on the other hand, small-scale, irregular movements of water called "turbulence," together with molecular diffusion, give rise to local mixing of the wastes. Since the diffusive power of oceanic turbulence is very much greater than that of molecular diffusion, the diffusive processes are often called "turbulent diffusion."

The term "diffusion" is sometimes used ambiguously so that it can mean "dispersion." Strictly speaking, both "diffusion" and "turbulent diffusion" are equivalent to "dispersion" *only* when there is no spatial variation of mean velocity in the water body.

INITIAL DILUTION AND DIFFUSER DESIGN

When waste water is discharged into the ocean there are essentially three stages of dilution:

1. Initial jet or plume mixing, close to the points of discharge;
2. Development of a homogeneous sewage–seawater "cloud" (sewage field) as the initial jet or plume energy dies out (transition between 1 and 3); and
3. Dispersion of the sewage field as a whole due to turbulent diffusion and shear in the velocity field.

The first stage is effectively managed by the type of diffuser structure used and the depth of discharge. The third stage is a natural process that presently can be managed only by establishing the initial extent of the sewage field. It is common current practice to use a multiple-port diffuser to achieve high dilutions over the outfall and also to take advantage of favorable temperature stratification.

Deep-water outfalls with long diffusers having many ports (in the order of 500 or more) can produce dilutions in the range of 1:300, while single jet discharges in shallow water will provide an initial dilution of only 1:5 to 1:10. The increase in dilution by one or more orders of magnitude through use of a diffuser is highly desirable and is less expensive than building outfalls that are long enough to accomplish the same increase in dilution by natural oceanic dispersion.

Another often-desirable feature of the multiple-port diffuser is its ability to generate completely submerged waste fields where there is density stratification in the ambient water. Off the California coast, a temperature differential of only 2°C between the surface and a depth of 200 feet is sufficient to prevent sewage effluent from rising to the surface if a multiple-port diffuser is used. The many small jets entrain heavier bottom water and the rising plume becomes neutrally buoyant before reaching the surface.

In the actual design of an outfall, which will be unique for each site, the engineer can choose the location, depth, and length of the diffuser, and the number and size of its discharge ports. Receiving-water quality objectives have the major effect on the design. Design decisions can be made only in a systems context, taking into account the degree of treatment, outfall and diffuser hydraulics, the various stages of dilution, the decay rate of bacteria and other substances, and the density stratification and current regimes in the receiving-water body. Water quality objectives also have a major effect on the design. In some instances a submerged waste field may be desirable, while in areas such as estuaries it may not be.

PHYSICAL PROCESSES IN ESTUARIES

Circulation Patterns

Traditionally the term "estuary" has been applied to the lower reaches of a river into which seawater intrudes and mixes with fresh water from land drainage. More precisely, an estuary is a semi-enclosed coastal body of water which has a free connection with the open sea, and within which seawater is measurably diluted with fresh water derived from land drainage.

The dilution of seawater in estuaries provides the density gradients that produce the characteristic estuarine circulation patterns. The basic factors that determine the circulation patterns in estuaries are the fresh water inflow and tidal currents. The physical dimensions of the estuary and the Coriolis force play supplemental roles.

The fresh water will tend to flow seaward as a layer of low-salinity water overlying the denser salt water from the sea, with the density difference between fresh and salt waters tending to cause vertical stratification. On the other hand, tidal currents provide kinetic energy to the system so that mixing of the fresh and salt waters can occur through a part or all of the vertical column. The extent of this vertical mixing depends upon the relative intensities of the fresh water flow and the tidal currents.

Part of the tidal energy is converted into kinetic energy of turbulence, which causes mixing, and some of it increases the potential energy of the water column by vertical mixing. For a narrow estuary, the effect of the Coriolis force may be negligible, but in a wide estuary it can cause lateral variation in properties.

THE HIGHLY STRATIFIED OR SALT-WEDGE ESTUARY

The best known example of a salt-wedge estuary is the Mississippi River, where the ratio of river flow to tidal flow is relatively large and the ratio of width to depth is relatively small. Seawater enters as a wedge along the bottom, while the fresh water flows seaward on top. The extent of intrusion of the wedge depends upon the magnitude of the frictional drag between the layers. Thus, during low river flows in the Mississippi, the almost undiluted salt wedge extends upstream for over 100 miles; at high river flow the salt wedge manages to intrude only a mile or so above the mouth of the river.

Because of a highly stable stratification in the water column, comparatively little mixing occurs at the interface between the layers. Interfacial waves are generated at the upper boundary of the wedge, become unstable, and break up. The process by which some of the salt water is entrained into the upper layer is essentially irreversible. As a consequence, the salt content of the upper layer increases slowly as the water moves toward the sea; the seaward volume-transport is increased by the amount of seawater entrained. The loss of salt water from the wedge to the upper layer must be compensated for by a slow flow of water into the wedge from the sea. The small amount of friction between the layers maintains the interfacial slopes slightly downward in the estuary direction.

PARTIALLY MIXED ESTUARY

In this type of estuary, the main source of mixing is no longer the breaking of interfacial waves associated with velocity shear, but rather the turbulence resulting from tidal motion. If a tidal motion of moderate amplitude is introduced into a salt-wedge estuary, the tidal velocities give rise to random movements of water throughout the water column. The turbulent eddies tend to mix not only the salt water upward but also the fresh water down into the lower salty layer, and the salt wedge ceases to exist as a distinctly identifiable feature. In a partially mixed estuary, there is no marked interface, such as that in the salt-wedge estuary, though commonly at about mid-depth there is a layer in which the salinity increases more rapidly than in the near-surface or near-bottom layers. The content of salt increases noticeably toward the sea in both the surface and deeper layers.

This type of mixing adds a greater volume of salt water to the upper layer than is true in the case of a salt-wedge estuary. As a result, the net (nontidal) circulation pattern involves flow volumes many times the volume of fresh water inflow. The oscillatory tidal currents, which are relatively homogeneous from surface to near the bottom, are superimposed on the net circulation. The net outflow in the less saline upper layer may, for instance, be as much as 20 times the river flow with the compensating net up-estuary flow in the more saline lower layer being 19 times the river flow. There is also a small net upward motion which is at a maximum in the midlayer.

Since the main parameter controlling the circulation pattern is the ratio of the amplitude of tidal currents to the river flow, there is a wide range in the degree of stratification occurring in partially mixed estuaries. In some cases, the total increase in salinity from surface to bottom may be as much as 10 percent while in others it is less than 1 percent.

VERTICALLY HOMOGENEOUS ESTUARIES

If the tidal currents are very strong in relation to the river flow, the mixing introduced by the tidal motion completely overcomes the stability resulting from the fresh water inflow, and tends to produce uniform salinity from surface to bottom. There is, of course, a horizontal gradient of salinity—the salinity increasing from the head to the mouth of the estuary. In contrast to the apparent lack of a vertical gradient in salinity, vertical shears of velocity occur and important vertical fluxes of momentum are present. Estuaries of this type may be subdivided into two categories. These are defined in terms of the ratio of width to depth.

If the ratio of width to depth is sufficiently large, the Coriolis force gives rise to a lateral variation of flow and salinity. Net seaward flow of lower-salinity water occurs on one side, and a compensating flow of higher-salinity water occurs on the other. Large-scale horizontal mixing occurs between the counterflows. There is some evidence that the lower and relatively wide portions of Delaware Bay and of the Raritan Bay are of this type.

In certain estuaries where the ratio of width to depth is relatively small and the tidal currents are very strong in relation to the river flow, the direction of water movement is symmetrical about the longitudinal axis. Variations in velocity are associated with the flood and ebb of the tide, and when the velocities are averaged over tidal cycles the net flow is apparently seaward at all depths. Vertical shears in velocity may still be present. The only nontidal variation in salinity is the increase of salinity from the head to the mouth of the estuary. In fact, the estuary with cross-sectional homogeneity in salinity may not exist in a rigorous sense. Nevertheless, certain estuaries can be treated, to a first order of approximation, as sectionally homogeneous. Theoretical and observational evidence have led to some quantitative relationships describing the properties of estuaries. One such relationship is the compound tendency of gravitational convection to generate vertical stratification, and of vertical mixing from turbulent tidal currents to destroy this stratification. While this quantitative relationship appears to be consistent with available observational evidence, too few direct observations of circulation in estuaries have been made to permit conclusive judgment of whether such simple parameters are adequate for prediction of the vertical distribution of properties in all estuaries.

SOME VARIATIONS IN THE ESTUARINE SEQUENCE

The estuarine types so far described can be considered as representing discrete examples drawn from a continuous sequence of possible types. Estuaries actually vary constantly in their characteristics, and may change continuously from type to type as conditions alter. In addition, there are deviations from this main sequence which deserve some discussion.

Horizontal circulations dominate certain wide, shallow estuaries in which gravitational convection is very weak. In these cases horizontal advection and horizontal diffusion apparently govern the distribution of properties in the estuary.

In bar-built estuaries, or sounds, the inlet between the sound and the sea is usually very restricted. As a result, they have characteristics different from typical coastal plain estuaries. Relatively weak tidal currents, combined with the relatively great width and small depth of such a system, permit wind-driven currents and wind set-up to provide the major mechanism for moving and mixing these waters. No adequate dynamic description of this type of estuary has yet been developed.

Another variation in estuarine circulation patterns may occur in certain tributary embayments to a large estuarine system, as in the Chesapeake Bay area. For example, so little fresh water is supplied to Baltimore harbor by the Patapsco River that it is ineffective in contributing to the density distribution. The adjacent Chesapeake Bay is the prime source of water for the tributary embayment. In the harbor, vertical mixing tends to reduce the degree of vertical stratification. Thus, though the depth mean salinity does not vary significantly along the length of the harbor and is very nearly the same as that found in the adjacent Chesapeake Bay, the salinity of the harbor decreases toward the bay near the surface and increases toward the bay near the bottom. As a consequence, a three-layer pattern of flow results with water moving toward the harbor in both the surface and bottom layers and outward at mid-depth.

Recently, it has been proposed that estuaries be classified as a function of quantitative relationship among top, bottom, and mean salinity, net nontidal seaward velocity and freshwater velocity.

While these proposed relationships appear to be consistent with available observational evidence, too few direct observations have been made to evaluate their adequacy.

Movement and Dispersion of an Introduced Waste in Estuaries

The degree of concentration of wastes in the vicinity of the point of introduction depends upon the initial mixing of the waste; the objective is generally to attain as high a dilution of the waste as possible from the initial mixing process. After an initial dilution, the diluted patch is transported by the currents in the estuary and spreads both vertically and horizontally by turbulent diffusion. The oscillatory motion of the tide supplies the major portion of the turbulent energy which leads to turbulent diffusion. Near the side boundaries of the estuary, the current exhibits a strong lateral shear which produces a deformation of the pollutant patch. Irregularities in the shoreline, such as embayments and protruding points of land, give rise to further complicated deformations. These effects, when combined with the transverse diffusion due to turbulent eddies, lead to an effective longitudinal dispersion of the wastes in the estuary.

Vertical diffusion of wastes is another factor that results in an effective longitudinal dispersion. This is particularly true in a partially mixed estuary when a two-layer flow pattern prevails. Thus, waste introduced into the surface layer, in addition to participating in the oscillatory movement and mixing due to tidal currents, is carried initially with the net flow toward the sea. Turbulent diffusion leads not only to horizontal but also to vertical spread of the wastes, and they are thus also added to the deeper layers where a net flow is directed toward the head of the estuary. The velocity shear in the net circulation pattern then produces a deformation of the waste patch and, when combined with vertical diffusive processes, eventually gives rise to an effective longitudinal mixing.

A similar argument may be applied to the dispersion of wastes initially introduced into the bottom layers of a partially mixed estuary. In this case, a small net vertical movement of water increases vertical diffusion, and brings up the waste into the surface layer. Then the shear effect operates on the deformed waste patch to enhance the rate of dispersion. In either case, the wastes are ultimately flushed from the estuary into the sea owing to the net seaward flow of the estuary.

When waste material is introduced as a continuous flow, as from an outfall pipeline, it is spread downcurrent in the form of a plume by the prevailing tidal currents. During ebb tide, the plume extends down the estuary, and during flood tide the plume extends up the estuary. Each reversal of the tidal current results in a folding back of the spreading plume. However, the plume seldom folds exactly back on itself. Thus, there develops a widespread field of relatively low concentration upon which is superposed, on each tide, a relatively narrow new plume of higher concentration.

Basic Relationships for Dispersion in Estuaries

The local time rate of change of concentration of a conservative waste is controlled by the processes of advection and diffusion. Motion in an estuary is always turbulent in character, and it is not possible to describe adequately the instantaneous waste field. One must deal with some mean concentration. The method of averaging depends, among other factors, on how precisely one wishes to describe the concentration field of waste in time and space. Thus, the average may be taken over, say, 3 minutes in time and 10 meters in space when one is interested in the horizontal distribution of contaminant concentration during a tidal cycle; the same averaging mode is taken for the velocity field in this case. On the other hand, averages over one or more tidal cycles should be taken when dealing with long-term distribution of wastes. In this case, the pertinent advective velocity is the net flow of the estuary, while the tidal currents are included in the turbulent components contributing to diffusion.

For each of the estuarine types described, and for the mode of averaging specified, a different set of terms dominates the basic advection-diffusion equation. In a salt-wedge-type estuary, the long-term distribution of contaminant introduced in the upper layer is determined by longitudinal and vertical advection and, to a lesser degree, by vertical diffusion. In a partially mixed estuary, the concentration distribution is determined by longitudinal and vertical advection, and vertical diffusion. In a sectionally homogeneous estuary, the concentration distribution is determined by longitudinal advection and longitudinal diffusion.

If one's interest is for a short time, say a day or so, during which the wastes stay away from the side boundaries of the estuary, special solutions developed for the oceanic horizontal diffusion (in an infinite sea) can be used. In these solutions, the variation of horizontal eddy diffusivity in time or distance from the center of the diffusing volume is taken into consideration. Although these solutions were developed for a water body of infinite horizontal extent, an approximate approach to their utilization is possible if one assumes that the physical boundaries of the estuaries simply restrict the size of area within which dispersion can proceed. Under these conditions, the applicable solution can be obtained from the basic solution for infinite space by the method of reflection or the image method. The same basic approach can also be applied to the problem of fewer spatial dimensions. That is, a solution applicable to a laterally homogeneous estuary can be obtained by averaging the basic solution with respect to the lateral dimension.

Recently the importance of the velocity shear in ocean horizontal mixing has been recognized. The basic result of this concept is that an effective (and additional) longitudinal dispersion is produced by the combination of a gradient of velocity with diffusion in the same direction. Since the net circulation pattern in an estuary is associated with a well-defined shear, especially in a partially mixed estuary, the importance of shear in estuarine dispersion is obvious.

The "shear effect" has a different appearance depending on whether or not the region is effectively bounded. Because the substance is allowed to diffuse indefinitely in the direction of shear, the effect of shear on dispersion is far more marked for an unbounded region than for a bounded region. In an unbounded region the effective horizontal diffusivity due to the shear effect increases without limit, whereas the diffusivity becomes constant in a bounded region. The constant diffusivity for the bounded case can be expressed in terms of the velocity profile and functions derived from it. In other words, the idea of the shear effect enables one to evaluate the effective longitudinal diffusivity from the velocity profile associated with the net circulation pattern of the estuary. Lateral shears are far more important than vertical shears in contributing to the longitudinal spread of wastes in natural streams like rivers. In estuaries, perhaps, vertical shears may well be as important as lateral shears.

Another approach to the mathematical description of the distribution of wastes in an estuary has been developed for the purpose of determining the probable large-scale average distribution of concentration. No attempt is made in this method to understand the detailed mechanisms of the mixing processes. Only the broad mechanics of the movement of the wastes through the estuary and their ultimate discharge or flushing to the open ocean are considered. The approach employs the so-called box model, or segmented model, wherein the estuary is divided longitudinally by vertical segments within which complete mixing is assumed. Exchange coefficients are introduced to represent the turbulent and tidal mixing between adjacent segments. Another parameter allows for transport by the residual flow, if any. These exchange coefficients and transport rates are determined from tracer experiments or observed distributions of salinity and the knowledge of the fresh water inflow to the estuary above the section in question. The values of the exchange coefficients and the transport rates thus obtained are used to obtain a first-order prediction of the temporal and spatially averaged distribution of an introduced waste and the mean flushing time of the system.

The Application of Models to the Prediction of Dispersion of Wastes in Estuaries

The preceding discussion has involved a description of the physical processes which control the temporal and spatial distribution of concentration of an introduced waste in an estuary, as deduced from the basic advection-diffusion differential equations; covered also were the implications of the analytical solutions of highly simplified forms of these differential equations. These equations might be called analytical models of estuarine dispersion. It is theoretically possible to obtain numerical solutions to the basic hydrodynamic and kinematic equations using high-speed computers. The complexity of these equations is such, however, that to date it has not been possible to treat the complete transient-state equations even using the largest high-speed computers. However, numerical modeling of estuarine circulation, salinity distribution, and introduced wastes is developing rapidly, using models of considerably greater complexity than have been employed where analytical solutions were sought.

In numerical modeling of estuaries, the equations of motion are used to determine both the mean circulation and the tidal flow. To date, the form of the Navier-Stokes equations of motion, in which the local and field acceleration terms are neglected, has been used for the mean flow, and the classical wave equations for the computations of tidal flow, using the rise and fall of the tide at the entrance as input. The continuity equations for mass and salt must be solved simultaneously with the equations of motion to predict the temporal and spatial distribution of both the circulation and the salinity. Prediction of the distribution of concentration of a waste requires the simultaneous solution of the advection-diffusion equations with the equations of motion.

To date, most work with numerical models has involved one-dimensional models and branched one-dimensional models (pseudo two-dimensional models). Some work has been done with true two-dimensional horizontal models. In all cases to date, the equations employed in the numerical model still represent considerable simplification of the complete hydrodynamic and kinematic equations. In particular, transient-state solutions for the nontidal circulation will ultimately be required. To treat the majority of estuarine problems, the vertical dimension must be included, since the two-layered circulation pattern dominates many of the estuaries in this country.

The greatest problem in numerical modeling of the estuary is that the relationships between the diffusion parameters which enter the equations and the predictor variables such as tide, weather, and circulation are not known. There are two schools of thought on how to overcome this problem. One contends that if the field of motion is determined in sufficient detail in space and time, then the diffusion terms become relatively unimportant, and only rough estimates of the diffusion parameters are required. The second school advocates determining the pertinent relationships by which the diffusion parameters applicable to any desired averaging scale can be computed. The advantage of the first approach is that at least two-dimensional modeling is possible now, though at considerable cost in computer time, since very short intervals in time and space are involved. The advantage of the second approach is that, for situations requiring much less detail in time and space, less computer memory and computer time is required, and the possibility of extending the model to three spatial dimensions appears more favorable.

Hydraulic models of estuaries have been and are being used to study the physical dispersion of proposed waste discharges. A hydraulic model is inherently three-dimensional, and time variations in tidal input and fresh water inflow are readily included in the model. Properly verified hydraulic models of estuaries are capable of producing valid information on the physical dispersion of conservative waste materials introduced into the estuary at intermediate and large scales of averaging. There is some indication that the details of concentration distribution near the point of discharge may not be properly scaled in the model, as a result of overmixing at small scales—a problem which is associated with the roughness elements required for distorted hydraulic models.

Costs of construction of hydraulic models of major estuarine systems are relatively large. Once constructed, however, the model can be used for many purposes. Taking all factors into account, the cost of operating an existing hydraulic model to obtain information on physical dispersion of an introduced waste material is not significantly different from the costs of obtaining the same information from a numerical model using a high-speed computer.

It is doubtful if the costs of construction of a hydraulic model of an estuary can be justified on the basis of waste management alone; however, the use of existing hydraulic models, built for other purposes, does appear feasible. This is particularly true for problems where all three spatial dimensions are important, since proven alternate means of solution are not yet available.

PHYSICAL PROCESSES IN COASTAL AREAS

For the purposes of this section, the coastal area is defined as the region between the coastline and the 100-fathom curve. It does not include estuaries.

The oceans are the ultimate sink for all of the wastes caused by man and his activities that are not buried, spread on the land, or discharged into the atmosphere (some of which reaches the ocean via fallout).

The path is by river and estuary to the coastal areas and thence to the oceans, or by direct discharge to the coastal waters and ocean. All that is added to the system is ultimately discharged to the oceans except that which is removed by geological, biological, or chemical processes en route. Each part of the system has a specific capacity to receive wastes without undesirable effects. This capacity depends upon the distribution of the sources, the physical processes of advection and diffusion, and the geological, biological, and chemical processes of removal.

There are approximately 860,000 square miles of coastal area surrounding the United States. By region, 16 percent is off the Atlantic Coast, 15 percent off the Gulf Coast, 3 percent off the Pacific Coast, and 65 percent off the Alaskan Coast. Because of the narrowness of the coastal area off the Pacific Coast, i.e., the proximity of deep water to land, the physical aspects of waste management in this region are, for the most part, reduced to problems of proper location and methods of introduction of the wastes. Hence, they are covered in the section on Initial Dilution and Diffuser Design. Because of the high population and industrial density along the Atlantic Coast, and the requirement for proper waste management, waters over the continental shelf between Boston and Norfolk are discussed here, since this region illustrates physical processes that influence management in coastal areas.

Mean Motion

The motion in the ocean can be considered as a continuous spectrum, encompassing scales ranging from the molecular free path to ocean-wide circulation. The division between the part of the motion that is assigned to advective processes, and that which is assigned to non-advective, or diffusive, processes depends on how much temporal and spatial distribution of the introduced waste is required. For purposes of this discussion, circulation patterns that are shelf-wide in extent, and at least seasonal in persistence, are regarded as the mean motion; all others including tidal motion are considered to be dispersive.

With respect to surface circulation in coastal waters, the average flow is parallel to the shoreline, with the shore to the right, in the Northern Hemisphere. The average motion of the surface water is anticlockwise in a bay or gulf, and clockwise around a bank. The reverse is the case in the Southern Hemisphere. The most pronounced surface current tends to be located near the 100-fathom curve. It is also characteristic to find a second—somewhat shallower and less saline—band of current near the beach.

The driving mechanism for this surface circulation is, of course, the river inflow. Because the relatively fresher water being discharged by the estuaries is lighter than the offshore waters, a cross-shelf density gradient is set up which gives rise to a circulation pattern with outflow at the surface and inflow at depth. As a result of Coriolis force, the offshore flowing surface current is deflected to the right in the Northern Hemisphere producing the observed surface currents parallel to the shoreline with the land on the right-hand side. Near the bottom, about all that can be said with certainty is that, owing to bottom friction, the flow must have a component in the direction of the pressure gradient, which is onshore at that level. Unfortunately, very little can be said concerning circulation patterns at depths intermediate between the surface and bottom.

Continuity principles provide insight into the magnitude of the volume of coastal water that is exchanged with oceanic water as a result of river inflow. Continuity of mass at steady state requires only that the volume of water flowing seaward through a vertical cross section extending from, say, Nantucket to Cape Hatteras be equal to the river inflow over the same period of time. However, since salinity typically increases with depth everywhere in these coastal waters, continuity of salt requires the upper layers to discharge many times the volume of fresh water drainage. If one assigns a salinity of 33 parts per thousand to the water leaving the shelf and a salinity of 35 parts per thousand to the slope water inflow along the bottom, then continuity requires the offshore component to exceed by approximately 17 times the volume of fresh water drainage.

Since 50 percent of the total annual discharge of river water is concentrated in March, April, and May, the currents due to river runoff follow a seasonal behavior with the density gradients reaching a maximum in the summer months when the fresh water accumulation on the shelf is a maximum.

Although coastal currents, unlike oceanic currents, operate without the direct help of the wind, strong and prolonged winds do cause important variations. Accordingly, the winds are a modifying factor rather than a direct cause of circulation in coastal areas. Along the east coast of the United States, winds with a southerly component cause upwelling, and an associated horizontal flow directed offshore in the surface layers and onshore in the deeper layers. Winds with a northerly component result in downwelling with an associated reversal in the horizontal flow pattern. They are most effective in modifying the normal current patterns during the winter when the onshore-offshore density gradient is a minimum.

One of the most critical phenomena, from the standpoint of transport of waste material onto a beach from offshore, is associated with an abrupt reversal of a wind which has been blowing for a prolonged period of time, i.e., approximately 4 to 5 days, from a direction that has produced upwelling. The upwelling results in steadily increasing nearshore density at all levels, presumably as a result of the offshore transport of light surface water and a subsurface flow of cold dense water toward the coast. Eventually, the interface between the light surface water and the heavier inshore water slopes offshore and downward. When the stress of the wind is withdrawn as a result of a windshift, this density distribution is unstable and the cold inshore waters sink and run offshore under the light surface layers which in turn move toward the coast. Obviously, any waste material contained in the surface layers which participate in this "sloshing" motion will be carried directly to the beach.

Dispersion

In this discussion we regard all scales of motion smaller than a certain scale as producers of internal shear and mixing, and we lump them all together under the term "dispersion." As a result, it is sufficient to quantify the dispersive characteristics of a given environment in terms of statistical parameters. One such statistical parameter is the rate of spread of a patch of waste or tracer material. Another useful parameter is the rate of decrease of peak concentration in a tracer patch. The usual procedure is to release a known quantity of tracer material, such as rhodamine dye, and to measure its three-dimensional distribution as a function of time. The measurements are then fitted to a particular mixing model and the dispersive characteristics quantified according to the model selected.

The point to be made here is that it is possible to measure the dispersive characteristics of a given environment and to predict the effect on the environment of introducing a given waste, either as an instantaneous or a continuous source, at least for scales up to a month in time and 100 km in space. However, it is difficult if not impossible to quantify the large-scale mixing processes that lead to renewal of the shelf waters by oceanic water. That is, the amount of tracer material released is determined by the sensitivity of the measuring instruments, which at present levels would require a tremendous amount of tracer for such an experiment. These large-scale mixing processes are of great concern, however, since they determine the capacity of the region under discussion to assimilate wastes from the east coast megalopolis.

One approximation of the renewal rate of shelf waters was made by computing the mean residence time (defined as the ratio of the total volume of river water accumulated on the shelf to the rate of river flow) from salinity measurements by 856 hydrographic stations. The rate of river flow was obtained from U.S. Geological Survey Water Supply Paper 1051 and the accumulation of fresh water by converting the salinity measurements to fresh water fractions, f, using the relation

$$f \equiv \frac{S_b - S}{S_b}$$

where S is the salinity of the sample considered and S_b is the salinity of the undiluted sea water, which was taken to be 35 parts per thousand. The computed mean residence or flushing times are shown in Table 7.

The mean age of residence time may also be estimated by assuming that waste is released as a line source paralleling the coastline. Exchange is considered to take place entirely by diffusion in the offshore direction. The relationship between time scale and diffusion velocity was

TABLE 7 Flushing Times of the Continental Shelf between Cape Cod and Chesapeake Bay for an Average River Flow of 14.5 × 10^9 ft^3/Day[a]

Depth Range, Fathoms	Flushing Time, Days		
	April–June	July–September	October–March
0– 20	112	125	108
20– 30	127	156	130
30– 40	119	157	123
40– 50	62	83	62
50–100	52	91	56

[a]Data from Ketchum and Keen, 1955.

TABLE 8 Flushing Times of the Continental Shelf between Cape Cod and Chesapeake Bay Based on a One-Dimensional Diffusion Model

Depth, Fathoms	Flushing Time, Days
0– 20	158
20– 30	94
30– 40	101
40– 50	61
50–100	77

obtained from an analysis of all available tracer experiments conducted since 1961.

Table 8 lists the flushing times in days obtained from the above equation, utilizing very crude estimates of the widths of the shelf segments having the designated depth ranges. The agreement with Table 7 is rather good, considering the crudeness of the model.

Adequacy of Existing Knowledge as a Base for Prediction

With respect to coastal circulation, it is fair to say that the driving mechanisms are well understood, with the possible exception of the role of the offshore currents, such as the Gulf Stream, in removing large masses of coastal water by eddies or meanders. Unfortunately, data obtained by occasional surveys of the distribution of temperature and salinity in coastal waters are of limited value when used in classical theories of physical oceanography because of periodic variations in temperature and salinity due to tidal currents and internal waves. What information is available has been obtained primarily from drift bottle and sea bed drifter releases. If a proper understanding of coastal circulation is to be achieved, continuous observations are needed at well-selected points across the shelf.

It is now possible to quantify the dispersive characteristics of a given environment for scales up to a month in time and approximately 100 km in space. Lacking, however, is information regarding the large-scale mixing processes that lead to exchange of coastal and oceanic waters. This deficiency is due to sensitivity limitations in the measuring equipment for present tracers. Development of a more economical tracer and/or more sensitive detection equipment is urgently needed if knowledge in this important area is to be advanced.

FLOATABLES, PARTICULATES, AND NONCONSERVATIVE WASTE COMPONENTS

Thus far, wastes that are either colloidal or in solution and are conservative have been considered. Some wastes, however, consist of particulate matter which is denser than sea water, and hence has a finite settling velocity; sludge is an example. Other wastes such as garbage, trash, oil, and grease float. It is clear, therefore, that models developed for materials in solution require modification if they are to apply generally.

Floatables

Objects that are constrained to move in a horizontal plane because of their density are subject to various effects when placed in a three-dimensional field of motion. For example, in the case of an onshore surface current, the shoreward component of the velocity must vanish at some point before reaching shore and the water parcels must either sink or change direction, or both. In any case, the result is a line running roughly parallel to the shoreline where floating debris collects.

A recent theoretical study of this problem concerned a group of floatables in a circle around a point of divergence or convergence of the velocity field. The results indicate that the floatables either grow or shrink with time, depending on whether convergence or divergence is involved.

If a random motion is superimposed upon the velocity field, the floatables will also be subject to advection-diffusion. As a result, the group size will tend to grow against a collecting power of convergence toward the singularity.

Sinkables

Particulate matter which is denser than seawater sinks, of course, and eventually is mostly deposited on the ocean floor, except for the fraction removed by biological and/or chemical processes. Mathematically, the problem of particles that sink (or rise) is more tractable than is the problem of particles that are constrained to one level. The simplest approach is to add a term to the convective-diffusion equation which expresses the time rate of change of material contained in a unit volume due to sinking (or rising). The effect of shear on the horizontal spread of particles heavier than water has been examined. Although little work has been done on the practical aspects of this problem, the difficulties do not seem to be insurmountable.

Of major concern is the pollutional significance of the material, once deposited. Many areas of the continental shelf off the eastern seaboard support commercial populations of edible shellfish, primarily sea clams, that are capable of concentrating and holding bacteria, viruses, and toxic substances which they ingest from their environment. In addition, a sludge blanket is detrimental to their growth as they require a sand or gravel bottom. At the present time there are several areas off the east coast in use as disposal sites for sludge, acid waste, alkali waste, and inert industrial waste. More work is needed to evolve a satisfactory method for determining the size of the exclusion area around a dump area, so that adequate protection is provided to nearby shellfish growing areas.

Nonconservative Waste Components

For the case of dispersion of a nonconservative waste constituent, such as a biodegradable component or bacteria, the advection-diffusion equations must be modified to include the rate of biochemical alteration of the component, or the growth and "die-away" of the bacteria.

Studies of modifications to the diffusion equations that are necessary to account for bacterial "die-away" in the dispersing waste plume in the ocean from a sewage outfall have shown that the nonconservative processes may be much more important than the physical processes of dispersion in accounting for the concentration distribution of waste components such as bacteria.

PHYSICAL EFFECTS OF WASTE COMPONENTS ON RECEIVING WATERS

Waste waters commonly contain significant quantities of floatable and settleable materials. Settling and skimming the waste water removes most of the settleables and a portion of the floatables, but the raw sludge derived from such a process has to be disposed of in some manner. Much of the organic content of the raw sludge can be destroyed by anaerobic digestion and the digested sludge is frequently disposed of by pumping the material through the treated effluent outfall or a separate sludge outfall. If the treatment plant is located in an inland waterway, the sludge may be barged to the open sea for disposal.

If secondary treatment is employed, the sludge disposal problem is generally increased by the additional load of secondary sludge. The floatables passing through the primary plant may be only partially reduced by the secondary treatment process. However, secondary treatment is rarely justified ahead of long marine outfalls on the open coast because there are no significant oxygen or BOD problems in this location. Comparatively few quantitative studies have been conducted on the removal of floatables in primary or secondary treatments, due to the lack of adequate sampling and analytical techniques.

Particulate matter is not the only type of floatable material that may cause water surface or beach pollution. Other forms are surface films, scum, and foam. Special techniques are required to sample and analyze such materials, and these techniques have not been investigated or developed for pollution studies.

The barging or pumping of digested sludge through an outfall adds another dimension to the problem. In this case some of the material settles to the bottom and the nature of the dispersion and spreading of the material on the ocean floor depends on the mode of release, the rheological properties of the deposited material, the slope of the bottom, the bottom currents, etc. This subject has been studied to a limited degree in the past but much remains unknown, especially with regard to barging operations.

Classification and Properties of Floatables

Floatables appear in the form of films, lenses, scum, foam, and particulate matter or detritus. Films may be classed as duplex, monomolecular (monolayers), or something in between. A schematic diagram of the approximate relationships between the various types of film materials is shown in Figure 2. Particulate matter may be found associated with the films or with scum and foam.

Monolayers represent a stable system; duplex films generally do not. Spreading oils may form duplex films first before stabilizing as a monolayer, or they may spread directly as a monolayer, depending on the relationship between the various surface and interfacial free energies. A nonspreading oil resides at the surface as a surface lens but the material "vaporizes," albeit slowly, onto the surface and produces a monolayer. Thus, theoretically at least, all liquids produce monolayers on a clean water surface.

Monolayers may be formed by soluble substances as well as by immiscible liquids. Soluble films are formed best by those substances which

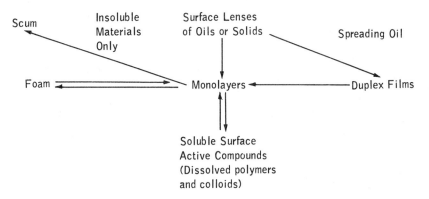

FIGURE 2 Approximate relationship between film materials.

are strongly adsorbed at the air-water interface. These three different types of monomolecular films—soluble, insoluble, and proteinaceous—possess distinctly different physical properties.

Scum may be formed by the compression of an insoluble film beyond its elastic limit. In that case one layer of insoluble material is piled on top of another, and if the material normally is in the solid state, scum is formed. Foam is a mixture of air bubbles, water, and the surface active materials. Foams are not stable, but on occasion massive quantities of quasi-stable sea foam are formed naturally and blown onto the coast.

The quantities of material required to form a compact monolayer are relatively small. For example, approximately 1 mg of protein will form a compact film 1 m^2 in area. This does not necessarily imply that film materials are difficult to sample or measure physically because they are so extremely thin. In fact, extremely precise methods of measuring films on liquids have been developed by the physical chemists who have worked in this area.

It must be recognized that floatables also occur naturally in the ocean. Since 1950, natural sea slicks have been investigated extensively by oceanographers and it appears that such slicks have the general physical properties of insoluble mixed-component monolayers containing an abundance of long-chain fatty acids and their derivatives. The source of the film material is believed to be the oceanic plankton.

Formation of Slicks and Streaks

Sea slicks are a common feature of coastal waters and estuaries during periods of relatively calm weather. The film materials have the ability to dampen capillary waves (ripples) produced by a breeze having a velocity in excess of 2 mph. Thus, a smooth or slick area is produced which contrasts visibly with adjacent cleaner waters.

Patches of film material may collect naturally if the substances making up the film have intermolecular cohesion. In general, floatables of all types tend to collect over the regions of convergences or descending water. Convergences are produced by many different types of natural surface circulation phenomena, including convection cells of various types and internal waves. The means by which the surface active materials collect at the water surface, or are transported along the air–water interface, are not well understood; it is known, however, that natural mechanisms exist for concentrating surface active materials of all types once they reach the surface. Thus, floatables generally do not stay scattered randomly over a water surface, but rather tend to collect and concentrate into slicks or streaks.

Apparently, slight quantities of floatables of waste-water origin may cause appreciable slick formation because (a) the materials are concentrated into a two-dimensional plane at the air–water interface and (b) they are concentrated still further into slicks or windrows by natural surface circulation phenomena.

Floatables

Investigations of surface materials of municipal waste-water origin have been extremely scarce. The results of the very few actually conducted on floatables tend to indicate that they may be blown onto beaches in objectionable quantities. For example, it has been found that most harbors and many beaches in California are polluted with oils, with the source of the oil being ascribed to natural submarine oil seeps, accidental oil spills, and industrial waste waters rather than municipal wastes. Investigators in Australia and South Africa have found strong correlations between grease content in beach sands and coliform counts in the offshore waters. Visible oil slicks appear to be a more sensitive indicator of oil spillage than any chemical extraction of bulk water samples.

Few investigators have attempted to sample and analyze the materials actually present on a water surface. Rather they have generally followed the indirect procedure of measuring the accumulation of grease or oil on beach sands, buoys, rocks, etc. Such procedures are useful in showing the effects of floating waste materials, but do not permit a complete assessment of the problem. In the few instances where the surface waters were sampled directly, the results tend to indicate that the amount of surface material may be a problem. Approximately 0.35 mg/liter of extractable substances were found in samples collected 1 mile downcurrent from the diffuser of a large marine outfall installation. The samples were collected by dipping a container into the water. At the time of sampling, the sewage plume was submerged and no evidence of its existence was visible on the surface.

In another study, an average of approximately 17 gm of extractable materials was collected per square meter of water surface area. The sample was collected by skimming the top 20 cm of the water within a region of 5,000 m^2 but most of the sample was collected within the confines of a surfacing sewage plume.

The only quantitative *in situ* measures of film pressures readily available were made in 1936. It was found that significant film pressures can be measured in most estuaries and within the vicinity of marine outfalls.

Floating particulate matter has been sampled and analyzed more extensively than film materials, primarily because the larger particles often can be identified visually and sampling techniques are simpler. Results of such studies have indicated the presence of appreciable quantities of particles of sewage origin as well as natural detritus.

TABLE 9 Status of Information on Jet and Plume Mixing

Type of Discharge and Environment	Round Source	Line or Slot Source or Multiport Diffuser
Continuous Discharge in Uniform Environment		
Simple buoyant plume	A^a	A
Submerged buoyant jet, horizontal (or sinking jet)	A	B
Submerged buoyant jet, directed upward (or sinking jet directed downward)	A	B
Submerged sinking jet directed upward (or buoyant jet downward)	B^b	B
Surface buoyant jet (e.g., hot water)	B	–
Continuous Discharge in Stratified Environment		
Simple buoyant plume in linearly stratified environment	A	B
Simple buoyant plume in irregularly stratified environment	C^c	C

TABLE 9 (Continued)

Type of Discharge and Environment	Round Source	Line or Slot Source or Multiport Diffuser
Horizontal, inclined, or vertical (upward) buoyant jet in linearly stratified environment	A	B
Horizontal, inclined, or vertical (upward) buoyant jet in irregularly stratified environment	B	C
Downward (inclined or vertical) buoyant jet in linear or irregularly stratified environment	C	C
Continuous Discharge in Ambient Currents		
Buoyant jet, vertical into uniform current	A	C
Buoyant jet, transverse into a current (uniform or stratified) (surface or subsurface discharge)	C	C
Sudden and Unsteady Releases (as from barges)		
All cases	C	C

[a]"A" indicates that good hydrodynamic models have been developed for gross behavior, and have been checked by experiments; future research is needed only to clarify details of velocity and concentration profiles and to determine more reliable entrainment or spreading coefficients.
[b]"B" indicates that hydrodynamic models have been proposed but not checked adequately by experiment; future research should include such laboratory experiments, and revision of models as required.
[c]"C" indicates that both theory and experiments are lacking. In all cases, good field observations have been extremely sparse and are vitally needed.

SUMMARY OF STATE OF KNOWLEDGE OF PHYSICAL PROCESSES

Initial Dilution and Diffuser Design

JET AND PLUME MIXING

The analysis of initial mixing over an outfall is based on the hydrodynamic analysis of jets and plumes. Much has been accomplished in this field in the last 15 years. Some of the more complicated problems require further attention. The state of knowledge in this area is indicated in Table 9.

DIFFUSER DESIGN

Present knowledge is inadequate to predict the jet pattern over a diffuser composed of many individual elements. We cannot determine when a series of jets act individually or collectively as a line source. If the ports are too far apart relative to depth, the dilution is not the maximum obtainable; however, there is no point in placing the ports closer together than is required to approximate a line source.

ESTABLISHMENT OF SEWAGE FIELD

The transitional stage between initial jet mixing and natural oceanic diffusion has been studied very little. No adequate way is presently known to predict the thickness of a sewage field or its average concentration as it moves away from the outfall. If the currents are weak, a thick sewage field can build up over the outfall thereby reducing the dilutions predicted by buoyant jet analysis. The thickness of the field must be such as to provide adequate advection to maintain steady state by the combined action of the ocean current and the accelerated transverse advection due to density differences.

WASTE HEAT DISCHARGES

When heated water is discharged from thermal power plants, the initial mixing problems involve the same type of hydrodynamics as buoyant jet problems. The differences are in scale and boundary conditions. Diffusers, for example, are not yet used in coastal waters, although they probably will be if higher initial dilutions are required.

Physical Processes in Estuaries

1. Broad understanding exists of the circulation patterns and exchange processes which occur in estuaries.

2. Most estuaries have circulations that are essentially two dimensional, either in the vertical or horizontal plane. In exceptional cases, a one-dimensional circulation is possible. For some in waste-management situations, the full three-dimensional nature of estuarine circulation may become important.

3. Dispersion of waste material occurs by both advective and diffusive processes. It is essential to recognize this difference in order to relate expected behavior to externally applied conditions. That is, the advective and diffusive processes depend upon different combinations of external parameters.

4. Diffusive processes should be restricted to those cases in which

there is a net flux of a constituent down its gradient without a net flux of water. In estuarine work, the effect of tidal motions may profitably be considered as either advective or diffusive depending on the particular problem at hand.

5. Numerical techniques which are rapidly becoming available permit integration of the differential equations describing two-dimensional circulations. However, numerical techniques for the vertical circulation and its complete three-dimensional circulation require development. The limiting factor in using these techniques derives from a lack of knowledge of the behavior of the appropriate frictional and diffusive terms as functions of the other parameters of the problem. In any case, one-dimensional and box models suffer from a similar limitation, in addition to the others imposed because of their oversimplification of the natural system.

6. Knowledge of the effects of turbulent, frictional, and diffusive terms is greatest for coastal plain, salt wedge, and shallow-wide estuaries. Less is known about the magnitude of these terms in fiords where their variation with stability and shear in the vertical direction plays a major role in the circulation.

7. The effect of wind on estuarine circulation and generation of turbulence has received little attention, but warrants substantial investigation.

8. In some portions of estuaries with weak wind and river-driven circulations, changes in external density differences may be an important driving force for the resulting circulation.

9. The physical processes that occur in estuaries not only play an important role in the distribution of waste material but also have an important influence on the chemistry, geology, morphology, and biology of estuaries.

Turbulent (Eddy) Flux

The idea of shear, or advective, diffusion has advanced to such an extent that one can evaluate the dispersion coefficient, provided the advective velocity field and the turbulent (eddy) fluxes are known. For an understanding of the physical processes of mixing in estuaries and coastal waters, knowledge is required of the horizontal and vertical eddy fluxes in relation to environmental parameters such as wind.

A considerable body of tracer diffusion data has been accumulated over the last 10 years or so, on which some empirical values of dispersion can be based to compare with theories. However, these data usually are obtained in the upper mixed layer of the sea above the thermocline. Because of operational difficulties, only a very few tracer experiments have been made in subsurface waters. These few experiments suggest that the rate of horizontal dispersion in these waters may be smaller by an order of magnitude than in the upper mixed layer.

Physical Processes in Coastal Areas

With respect to coastal circulation, it can be said that the driving mechanisms are well understood, with the possible exception of the role of offshore currents, such as the Gulf Stream, in removing large masses of coastal water by eddies or meanders. Occasional surveys of the distribution of temperature and salinity in coastal waters are of limited value when applied with classical theories of physical oceanography. The periodic variations in temperature and salinity due to tidal currents and internal waves require continuous observations at well selected points across the shelf.

It is difficult, if not impossible, at present to quantify the large-scale processes that lead to removal of shelf waters by oceanic water. These large-scale processes are of great concern, however, since they determine the overall capacity of a region to assimilate waste. For example, the coastal area between Cape Hatteras and Nantucket has a certain limited capacity to receive the wastes from the east coast megalopolis. The determination of this capacity for all coastal areas is a necessary objective. One reason our present knowledge is inadequate is because of the limitations in the sensitivity of available measuring equipment for existing nonradioisotopic tracers.

Decay of Nonconservative Constituents as Related to Physical Factors

Little is known concerning the relative significance of the various physical, chemical, and biological processes affecting the net dispersion of the nonconservative constituents of waste waters discharged into the marine and estuarine environment. For engineering requirements, quantitative relationships between the parameters describing these processes and related environmental factors are not adequate.

Interactions between Floatable and Settleable Components of Wastes and Physical Factors

Waste waters commonly contain significant quantities of floatable and settleable materials. The methods of waste-water release employed, i.e., barging or outfall, may influence greatly the spread and distribution of the settleable materials in coastal or estuarine waters, as well as the prevalence and character of the materials appearing as floatables at the air–water interface. Relatively little is known about the physical transport and spreading of settleable materials in the marine environment, especially when the materials are released from barges, and almost nothing is known about the prevalence or nature of the floatables derived from waste-water sources. The effective design of waste-water outfalls in part, and sludge barging operations in full, requires information on the initial and ultimate fates of these materials.

Chapter 4

Chemical Factors

The increase in human population and the demand for a higher standard of living are causing serious stresses on the various ecosystems. An effort to manage these stresses rationally will enhance their advantageous aspects, and diminish their ability to deteriorate not only man's local aquatic environment but the entire world ocean.

The illusion of the inexhaustibility of the ocean arises from its great total volume and the large flows of water and chemical substances into and out of it. Of interest are the following data descriptive of the ocean:

Total Volume	1.37×10^9 km^3
Surface Area	0.361×10^9 km^2
	(=71 percent of the area of the globe)
Annual Evaporation	124,000 km^3
Annual Reprecipitation	87,000 km^3
Annual Run-Off from Land	37,000 km^3 (37×10^{12} tons)
	(1/38,000 of the volume of the ocean)
Annual Flow of Dissolved Minerals into the Ocean	4.4×10^9 tons

It can be assumed that all of the naturally occurring elements are present in the ocean to some extent. For all but about a dozen, the amounts have been directly or indirectly determined. The tonnages are large, even for those present only in very low concentrations. Would-be inventors and others have been intrigued by the fact that the oceans contain 15 million metric tons of gold, but the concentration is only 11 nanograms per liter. The amount of radium in the ocean is 85 tons, present at a concentration of 60 mg per km^3 (0.060 picogram per liter).

The chemical composition of the ocean tends to be stabilized not only by its great diluting capacity, but also by the fact that the concentrations of most constituents are in a near steady-state condition, being removed by various processes at about the same rate that they are being added. In spite of these stabilizing factors, it is possible for man to affect the ocean in significant ways. The effect of a changed rate of flow of a substance into the ocean as a result of man's activities cannot be judged by what would happen to the average oceanic composition; local effects are more important than the average effect.

The ocean is well stratified, and mixing is slow. A surface layer a few hundred meters thick may accumulate man-made wastes that are diluted by deeper water only very slowly. This appears to have happened in the case of lead, which passes from automobile exhausts into the atmosphere and then passes in part into the ocean. The lead concentration in the upper layers of the oceans is now several times the average lead concentration in the total ocean.

Discharges near land, where the water is shallow, can have significant effects on the local ecology. The water overlying the continental shelves is only 8 percent of the total volume of the ocean, and circulation is often limited. In some areas, essentially separated from oceanic current patterns, nutrient elements and organic materials added by man can cause serious reductions of oxygen levels. Even a sea as large as the Baltic has been reported to be affected in this way.

Estuaries and small bays are, of course, even more easily affected. Those byproducts of man's activities that reach the ocean travel mostly by way of rivers, making the estuaries especially vulnerable.

The tendency of the ocean to maintain relatively constant concentrations of many elements, due to removal processes nearly equaling the rate of natural additions, does not insure that those concentrations will remain the same under the impact of man's activities. An increased rate of addition of some elements may result in an increase of their concentration if the reactions by which they are removed are relatively slow. The case of lead from leaded gasoline is an example.

Even if the total dilution capacity of the ocean does reduce a waste to an extremely low concentration, its effects will not necessarily be negligible. For example, marine organisms have the phenomenal ability to increase the concentration of essential as well as other trace elements in their body tissue to 10,000 times or more that in the water. Physicochemical, as well as biological, processes can nullify the diluting capacity of the ocean and other natural bodies of water. These processes include precipitation, flocculation, ion exchange, sorption, and others. No doubt the physicochemical and the biological accumulating processes often act together to produce results which under some conditions can be serious.

The concentration of radioactive strontium in some estuarial areas is illustrative. This effect was curbed by the cessation of atmospheric nuclear explosions. Mercury present in an industrial waste was accumulated in edible fish and shellfish in Minamata Bay in Japan in amounts that caused the death of 40 human beings, as well as some domestic animals. Concentrating insecticides of the DDT group in successive steps of food chains is related to disease or reproductive failure in fishes and birds.

The fact that man can cause significant, and frequently harmful, effects in the marine environment does not mean that man must cease all activities that add unusual constituents to marine waters or that materially change the natural fluxes of the elements. It would be impossible for man to carry on normal activities without affecting his environment, just as other organisms do. He has been drastically altering his terrestrial home from prehistoric times. With the present population and level of activity, further changes are to be expected of which those due to the handling of wastes are only one aspect. The appropriate approach to the problem of waste handling is scientifically based management. Wastes should be managed in ways that minimize harmful effects and maximize those that are beneficial.

Optimum waste management can be attained only through adequate knowledge of the ecological effects of the wastes under a wide range of conditions. One cannot imagine a time when all will be known about the ecological effects of wastes, but it is important that more information be acquired than is known at present. Biological, chemical, and physical effects must be considered, and plans must be made for various programs of observation to monitor the effects of wastes that are discharged. The primary purpose of this chapter is to outline the areas of needed research on the chemistry of the ocean. The term "chemistry" is here used comprehensively, and includes general organic, inorganic, and analytical chemistry, and all physicochemical, geochemical, and biochemical processes.

DISCUSSION OF RECOMMENDED FIELDS OF RESEARCH

The recognized important areas of information can be divided into five main categories:

1. Inorganic chemical processes.
2. Chemistry of particles and processes in sediments.
3. Nutrient chemistry and biochemical changes.
4. Chemistry of specific wastes.
5. Chemical consequences of man's physical activities.

Presented first are background discussions of these categories. The discussions are necessarily very sketchy, but provide the basis for an understanding of the significance of knowledge in the various areas.

Inorganic Chemical Processes

Much information has been obtained in recent years on the concentrations of trace elements in the ocean; at present, however, our knowledge is far short of adequate for even such a significant area as the concentration variation with depth, temperature, and biological activity.

Not much is known about the forms in which many of these elements exist, and their dependence on such factors as pH and redox potential. Chemical speciation may be of overshadowing importance with respect to biological effects. One example among many that might be given is the ionic state of zinc. At pH 8.3, the equilibrium concentrations of the inorganic forms Zn^{++} and $ZnOH^+$ should be about equal. The $ZnOH^+$ would be more highly adsorbed on particles than the divalent form. Determinations of zinc on filtered ($0.45\,\mu$) and unfiltered water off Acapulco, Mexico, showed that almost all the zinc was present in particles. The biological consequence of this phenomenon cannot presently be evaluated, but possibly the zinc could be stripped from the water to concentrations below those essential for life by adsorption on particles, particularly if there were any increase in the pH. As the pH decreases, the divalent ion becomes more and more predominant, and thus the biological availability may be altered.

The metals in solution in the ocean are complexed by organic as well as inorganic ligands, with important consequences for the biological-biochemical role that metals play in the ocean. The biological behavior of a strongly complexed metal ion is quite different from the role of the metal as a "free" ion. One need only cite the cases of magnesium in chlorophyll, copper in hemocyanin, and cobalt in vitamin B_{12} to make the point. The mercury in Minamata Bay, Japan, apparently owed its toxicity to the fact that it was present in an organo-complex form. There are several biological phenomena in the ocean which may be related to the reduced or enhanced availability of metals, which influences both toxic and nutrient effects.

Knowledge of the forms and transformations of the elements, particularly as they control the steady-state condition of the concentrations of these elements, will enable us to predict the effects of changes in the fluxes of the elements caused by man's activities.

Chemistry of Particles and of Processes in Sediments

The chemistry of the ocean is not simple solution chemistry. Phase changes are not accounted for entirely by solubility products and dissociation constants as ordinarily used. Solid phases, both organic and inorganic, and both suspended and on the bottom, play a dominant role. The processes of adsorption, biological accumulation, and transport, flocculation, solid solution formation, and solid phase reactions are of major importance in maintaining the intricate balances of ocean chemistry. It is now generally agreed that metatheses involving clays in the bottom are important factors influencing the pH of the ocean and hence the carbon dioxide content of both the ocean and the atmosphere. Because of the slowness of some of these reactions, and the slowness of the mixing of the ocean, it may take many millennia for a new steady state to be reached after a change in the inputs. If more were known about the reactions of particles and bottom sediments, as well as the solution chemistry involved, it probably would be possible to estimate the effects of such changes.

Oxidation–reduction reactions in the sediments, and sometimes in the water, have important effects. The oxygen resources of the ocean are not always sufficient to maintain aerobic conditions, especially at the bottom, where dead organic matter collects.

The advent of reducing conditions, signaled by a large drop of the oxidation reduction potential, usually permits the bacterial reduction of sulfate to sulfide as the first conspicuous manifestation. In the water in contact with an anoxic bottom, there are larger than usual concentrations of organic matter, PO_4, NH_4, and silicates; nitrate is absent; and the pH is lowered.

Changes occur in the oxidation states of metals. For example, the amount of iron in equilibrium solution in oxygen-rich waters is probably controlled mainly by the solubility of ferric hydroxide, which is considerably less soluble than ferrous sulfide. The latter, apparently, sometimes controls the concentrations of soluble iron over anoxic bottoms, as in Lake Nitinak on Vancouver Island. Manganese, too, goes into solution when converted to the divalent form; its sulfide is relatively soluble. By contrast, the concentrations of copper in solution are reduced to vanishingly low levels in the presence of sulfide, because of the low solubility of copper sulfide.

When an area of bottom has been anoxic for some time, the accumulated organic matter, sulfide, and metals in lower oxidation states act as a reservoir of oxygen demand. If the source of organic matter responsible for anoxia is stopped, it may be a long time before an aerobic condition is re-established.

Such are a few of the effects of anoxia that have been little explored or have been determined only qualitatively. Some anoxic bottoms are produced by man's activities, especially in estuaries, bays, and fiords, and in some instances even in local areas of open coast.

Nutrient Chemistry and Biochemical Changes

The flux of nutrient elements is extremely important in determining the condition of the ocean. If the sea is to produce marine animals, there must be an adequate supply of nitrogen, phosphate, silica, and certain minor elements to support the required growth of phytoplankton. Too great an abundance of nutrients, however, causes a vegetative overgrowth, producing noxious conditions at the surface and anoxic regions in the deeper waters.

Man influences the flux of nutrient elements in the ocean, both by adding these elements through waste-water discharges and land drainage, and by the removal of fish. Man removes about 6×10^7 tons per year of seafoods from the ocean, probably containing 2×10^5 tons of phosphorus. It has been estimated that man's activities are adding 1.8×10^5 tons of phosphorus a year. While this flux is negligible in comparison to the global preserve of about 10^{11} tons, the significance of these quantities cannot be weighed until one knows more about the cycles of phosphorus within the ocean, particularly the processes of precipitation and return to solution. It has been found that man has no appreciable effect on the supply of nutrients for marine life processes, taking the ocean as a whole, though he may cause marked local effects.

Dissolved organic matter is present in the ocean at all depths at a concentration estimated to average about 0.5 mg/liter. This quantity is much greater than the annual production of organic matter by photosynthesis. Many classes of substances or individual compounds have been measured: fatty acids, other organic acids, fatty esters, complex lipids, hydrocarbons, free and combined amino acids, carbohydrates, vitamins, and toxins. Still, a complete characterization of this organic fraction has hardly begun. Even more complex is the particulate organic matter. It needs at least proximate classification. Its sources, fate, and biological significance should be understood.

Studies of trace elements, nutrients, and organic compounds will assist in the development of understanding of plankton blooms.

The large influence of minute traces of certain elements depends upon biological concentration processes. Often these processes are compounded by successive steps. Animals, starting with trace elements already concentrated by plant life, may build the concentrations to higher levels.

Not only inorganic elements, but some organic compounds, if resistant to decomposition, may be concentrated biologically. The accumulation of chlorinated hydrocarbon insecticides is spectacular, building up in successive steps to levels sometimes a million times as great as that in ambient water.

The mechanism of accumulation of the chlorinated hydrocarbons appears to be quite simple, being accounted for by the high solubility of the compounds in body fats. The processes by which other trace substances are extracted from the water are obscure. Where the accumulation is essential to a life process, knowledge of the reaction mechanisms may be relatively academic, but where these same mechanisms lead to the pickup of other elements of analogous properties (for example, strontium instead of calcium), they may be important for understanding the cycles of these other elements. Knowledge of the reaction mechanisms may help to explain, or to predict, the effects of changes in the ecology.

Organisms respond at various levels to chemical changes in their environments. The highest level of response is death. The biota of a region may be profoundly affected, however, at concentrations far below those causing death. Reduced vitality or growth, reproductive failure, loss of forage, or interference with the functioning of chemosensitive receptors can cause a species to decline, or may allow another species to forge ahead because of reduced competition. It has long been recognized that the concept of median lethal dosages, while serv-ing useful purposes, is an inadequate indicator of tolerable levels of a chemical substance in the environment. Some progress has been made on the effects of sublethal concentrations of toxicants, but what has been done is only a preliminary effort in view of the magnitude of this problem. Knowledge of sublethal effects is extremely important in waste disposal management.

The Chemistry of Specific Wastes

An increasing variety of synthetic organic products is becoming available for industrial and consumer use. A large part of these products are transformed by natural processes into the elementary forms common in nature. A few, however, are so resistant to change that their presence in solid and liquid wastes can have far-reaching ecological effects.

Only a few years ago, the chlorinated hydrocarbon insecticides were considered ideal because of their high toxicities to insects and their relative harmlessness to other species. Recently available information, however, indicates some serious complications.

Absolute security against harmful effects can never be assured, but to wisely protect the environment, more knowledge of the problems existing today will be required. What are the fluxes of potentially hazardous substances produced by man? How can the degree of hazard be determined? How shall the effects of products being developed be predicted?

The environment may be altered, not merely by the addition of man-made chemicals, but by gross changes in the flux of compounds of natural origin. The outstanding example of this is the spillage of petroleum in the ocean. More must be known about the effects and the fate of oil spilled into the water. How is the oil acted upon by physical and chemical forces? What effects does it have on such processes as gaseous exchange at the surface, or oxygen balance?

Chemical Consequences of Man's Physical Activities

Man can alter the chemical state of marine waters by ways other than addition or subtraction of chemical species. He may, for example, affect the transparency of the water by contributions of sediment from dredging, land erosion, terrestrial and marine mining, etc. Photosynthesis may, therefore, be affected, restricting plankton activity to upper levels and confining attached algal growths to shallow water, thus changing the oxygen distribution. The steady-state reactions of the bottom sediments with the overlying water may be altered by changes in the kinds and quantities of sediments being deposited. By stopping or increasing the normal flow of sediments into the ocean, he can alter the local benthic ecology.

RESEARCH PRESERVES

Because of the enormous complexity of the physical, chemical, and biological interactions in marine ecosystems, there should be a scheme of ecological classification that will allow first approximations of the impact of specific types of wastes on natural coastal systems. Such classifications may require new geochemical surveys of such systems, but in large part they could be erected from present knowledge.

Classification of coastal systems should be used to select and set aside type preserves for experimental use such as stressing the environment to determine the effects of the stress and the rate of recovery of the system when the stress is removed. Such study areas would allow the carrying out of experiments that might not be permissible in areas not so set aside and would minimize the intrusion of other human influences. The stresses applied might include the addition of growth-suppressing substances, nutrient substances, heat, etc. The study areas would have to be provided with adequate laboratory facilities for intensive investigation and manipulation.

Types of systems to be set aside should include tropical, temperate, and boreal systems such as open coastal areas, salt marsh estuaries, tidal estuaries, salt-wedge estuaries, mangrove swamps, and fiords. They should encompass systems of varying depth, size, and geomorphology to permit collection of data from which to construct models useful for coastal waste management. This establishment of preserve areas is considered to be a matter requiring immediate attention.

Chapter 5

Biological Effects

The communities of marine organisms found today are the evolutionary consequences of biological adaptations (a) to the magnitude and variability of the physical and chemical parameters normally encountered in the environment, and (b) to the past interactions of their biological constituents. All communities are fragile in the sense that they are susceptible to stresses that are not part of their historic experience. Many of the substances entering the sea today as wastes are clearly not part of this experience. Depending on the level of stress they impose, such substances can reduce population sizes, exterminate species, and even eliminate entire biotas.

The wastes that find their way to the sea, either directly or indirectly, inevitably cause biological effects because of the increasing concentrations of naturally occurring substances and, more critically, the introduction of exotic materials. Since the quantity and types of wastes that are being dispersed into the sea are increasing, it is of major importance that the ability to predict the consequences of their presence and to find ways to control these effects be improved markedly.

Present abilities, insights, and state of knowledge permit predictions with varying levels of confidence. However, one prediction of paramount importance that can be made with a very high level of certainty is that if the degradation of the environment continues at its presently accelerating rate, the result will be catastrophic, primarily to the biological systems involved.

This prediction has its roots in the nature of dynamic society. The increasing population and the spawning of new products is increasing per capita waste production. Some of these products are clearly the creation of artificially induced demands. The effects of many products and their wastes on the marine environment are vaguely comprehended. Yet, the legitimate needs of an enlarging population impose more massive requirements on our resources and greater waste management problems.

If the disposal of wastes can be kept within the assimilative capacity of the coastal waters, the benefits will be enormous. Achievement of this goal will require comprehension of the biological communities and processes of these areas, and establishment of firm limits on waste disposal to keep it within the capacity of the system to assimilate.

GENERAL ASPECTS

Quantitative Analysis of the Biota

Since the biological effects of waste disposal are of major concern, it is of particular importance that these effects be evaluated critically at the level both of single species and of community response. The evaluations must be based on quantitative studies of the biota in the field. Studies are needed to obtain point estimates of parameters describing both the dynamics and the structure of single species populations, and of communities composed of many species. Unless measures of variability are associated with the estimates of each parameter, such estimates are of little use individually and objective comparisons between them cannot be made. In the case of biological studies, added to the random variability expected of samples of the populations is the variability due to activities and distribution patterns of the organisms. This variability may be expressed as either broader confidence limits or as bias in the estimate. The first arises most frequently from the fact that organisms are usually aggregated in time and space; the second, from selective collection based on the interaction between the sampling equipment and methods and the movements of the animals. To measure the precision of the estimates, it is necessary to take several replicate samples. These replicates should be as nearly identical as possible in all properties expected to affect the results (space, time, method, etc.).

Fundamental studies are greatly needed to separate and assess the various components of sampling variability. These studies should include: comparison of different methods and equipment in terms of their effectiveness when used on the same species and their interaction with species type when used on several species; and development and testing of new methods, equipment, and model studies by the application of various sampling schemes to synthetic populations with known, realistic properties.

Even if the components of variability are understood, it is inefficient, and often misleading, to obtain only a single set of samples. The sampling design (timing, position, number and type of samples) and the methods of data analysis interact and strongly influence each other. The question must be stated precisely so that an answer is possible; the sampling design must include some randomization; and the data analysis must be compatible. All factors should be decided on in advance. Any ecological study that does not meet these requirements is neither quantitative nor objective. Excellent presentations of the theoretical and practical aspects of sampling design and data analysis are available in the literature. Because they have seldom been considered carefully, it is often difficult or impossible to interpret the results of past surveys.

A variety of hardware and techniques is available for sampling microplankton and zooplankton populations at point locations, over distances, and at given depths, but most of these methods provide only a rough estimate of sample volume. Accurate sample volumes are essential to accurate estimates of concentrations.

Benthic sampling equipment now available does not always allow easy determination of sample area. There are also problems with penetration of the substrate and winnowing of material as it is being brought to the surface.

The biological components which are the most difficult to sample quantitatively are the larger invertebrates and fish. Trawls produce a shock wave as they are pulled through the water so that some mobile invertebrates and fish avoid them. New methods such as the helicopter-borne purse seine still have problems and limitations in practical application.

Assessment of Effects

It is now impossible to make quantitative predictions of the effect of waste disposal in the oceans, or even to measure the full effects of such introductions as have occurred. The complexities of the problem seem overwhelming, and include at least the following real difficulties:

1. Too little is known of the effects of the total marine environment on chemical processes, and of the effects of chemical form on entry and use in biochemical processes. Extensive bioassay of pesticides and other chemicals has demonstrated that each compound should be fully tested since there is only general prediction value even between related compounds.

2. The biological communities of the continental shelf are poorly known, with insufficient comprehension of species composition, energy flow, food chains, and environmental relationships. They vary so greatly around the coasts that generalized prediction is not appropriate.

3. Thousands of species may be involved, and each species is a biochemical and biological entity, sufficiently different from others to make transfer of predictions difficult or unreliable, even between rather closely related species.

4. Within species, response to toxicants has been shown to vary greatly with different stages in the life history, with previous nutrition, with season, and with unknown causes.

5. Toxic effects on single individuals are complex, and may involve different biochemical mechanisms at different concentration levels. Time–exposure–effects curves are frequently nonlinear, so that extrapolation may be difficult.

Since only partial assessment of toxicity is presently being considered, the magnitude of the limitations should be appreciated.

The conclusions that the coastal environment must be assessed for its useful capacities and that wastes must be managed to protect the quality of the valuable areas involved are unavoidable. Therefore, the question is: What are the best available methods of making limited but useful estimates of toxic effects, and how should they be applied?

FIELD STUDIES

Field methods for toxicity assessment in the coastal environment have the inherent disadvantages of high expense, long time schedules, and dynamic complexity in the area under study. They have the unique advantage of involving the actual site of effects. Field toxicity studies also have the principal advantages that they can disclose gross effects, can provide highly valuable clues for laboratory studies by indicating the ranges of environmental changes, species, and life history stages that are appropriate for further research.

Several types of field studies have been made, including before-and-after comparisons, and comparison of the receiving area with a so-called "control" site. The best design would include thorough pre-waste discharge comparison of the disposal site with a similar reference area, and adequate continuing observation of both after disposal began. There have been few, if any, adequate field studies of coastal waste disposal despite many partial efforts.

Methods of estimating community parameters such as standing stock, productivity and diversity indexes are straightforward once the basic data are obtained—numbers of organisms per species, biomass, diurnal dissolved oxygen changes, oxygen differences between light and dark bottles, and ^{14}C uptake. Classifying plants and animals requires skilled and experienced taxonomists. They, and those familiar with techniques for measuring productivity, are in short supply. The greatest inadequacy in this area appears to be the lack of trained and experienced personnel.

The application of measures of standing stock, productivity, and respiration to ecosystem response has been somewhat limited. It has led to models of succession, to models of energy flows within a biological community, and more recently to energy network diagrams.

The principal strength of these approaches is that any ecosystem may be described by the flow of energy through it, and its response to environmental changes or constraints such as tidal mixing, turbulence, nutrient cycling, organic waste inputs, toxic waste inputs, etc., may be simulated. This fact allows the consolidation of the many facets of ecological research and the work on biostimulation and toxicity bioassays.

The principal inadequacy of any type of model is that not enough data have been available at any one place to test the model rigorously, establish levels of response, or estimate inherent variability. Most models are incomplete because they do not take into account changes in the species composition of the ecosystem.

Diversity indexes have been applied successfully in certain areas to ascertain quantitative relations between the "health" of the benthic animal population and waste discharges. These indexes summarize community structure in a single value. Their magnitude is determined largely by the more abundant species and little affected by the rarer ones. As summaries, they lose information concerning the identity of the species involved and thus may obscure major changes in species composition. These changes are often indicative of changed conditions. The information can be retained if, in addition to the diversity index, a species list in rank order of abundances is given. More research is needed to determine the variability of these indices when based on replicate samples from the same community, and to define their relationship to specific ecological conditions.

Because the organisms that are successful in polluted habitats are generally those species that are tolerant of a wide range of conditions, indicator organisms will probably never be satisfactory as unique tests for areas of pollution. Though a vast amount of work has been done in Europe and in this country on attempting to relate specific biological response to waste concentrations, no workable and quantitative system exists today.

LABORATORY STUDIES

Laboratory methods are limited to small portions of the real ecosystem, and involve serious problems in achieving reasonable imitation of the natural environment (pressure, light, chemical composition of water, substrates, gas concentrations, bacterial populations, and turbulence or other waste movement). Perhaps this is the reason why so little useful data exist on marine toxicity and tolerances. On the other hand, such experiments can reduce the complexity of our views of the natural system, provide for adequate replication, and place the experimental organisms under the continuous observation of the investigator. They sometimes permit multivariate experiments, interaction between computerization and experimentation, and other useful tools in ways which are not applicable in the ocean. If properly designed, laboratory studies have potential for universal application which is rare in field studies.

Adequate attention must be given to several difficult aspects of experimental procedure. These include the selection of appropriate test species and life history stages, establishment of appropriate environmental conditions, acclimation of organisms, selection of experimental exposures to toxicants, monitoring for changes in resistance or tolerance, interpretation of the response obtained, and inclusion of both acute and chronic exposures.

Assessment of the effects of toxicity is obviously complex, and the difficulties are only suggested here. There is no alternative to the thorough studies suggested if the objective of useful prediction of the principal effects of coastal waste disposal is to be achieved.

Providing Guidance for Decisions

With the advent of the Water Quality Act of 1965, the individual states and the national government accelerated their programs of waste control and management. This action required the generation of water quality standards for the maximizing of beneficial uses of the receiving waters. For the development of standards, biologists, at an early stage, formed into multidisciplinary teams responsible for developing water quality criteria. They soon recognized that the assimilation of wastes in the receiving waters was one of the many important legitimate uses of the aquatic environment. It was required that these biologists propose guidelines that would allow for the assimilation of wastes and, at the same time, protect the many, often conflicting, uses of the water, including the protection of the biological community. It was found that, for the most part, if the biota could be protected, most other uses would be preserved. The biologist's role in the decision-making process of waste management is essential; he contributes to the establishment of receiving-water criteria which serve as guidelines for engineers in developing waste management processes and facilities.

The biologist participating in decision-making processes must have quantitative and qualitative descriptions of the wastes being discharged, as well as their variation in time and space within the receiving waters. Spatial distribution and characteristics of wastes in the environment are the keys to site selection for points of discharge. The discharge locations should be selected to minimize the impact of wastes.

In many cases, the biologist is asked to propose criteria for rather specific situations. For example, he may be requested to develop standards for an oyster fishery as it relates to the effects of cannery wastes or some other type of discharge. His knowledge of oyster biology and environmental requirements will serve to set guidelines for decisions on how much waste can be assimilated in the receiving water, and thus, in an indirect way, the degree of required treatment. It is emphasized that each situation must be evaluated in terms of the biota to be protected, and the type of wastes to be discharged, as well as the physical, chemical, and geological characteristics of the area. Years of experience have shown that no single set of techniques can be used for all situations. The biologist must adjust his investigations in accordance with the conditions at hand. Consequently, he must have a battery of tools and approaches available for his use. For future and present work, new investigative procedures must be developed, and old ones upgraded. A number of approaches must be used and the results from each must reinforce the others.

To aid the decision-making process, response variations resulting from natural phenomena must be isolated from biological responses brought about by waste management practices. Consequently, it is essential that studies be designed so that the appropriate statistics and relationships can be used. For example, statistical programs such as multiple regression, isolation of variance, factorial designs and many others have been extremely useful in environmental studies.

Research programs designed to expand the array of investigative tools are vitally needed. These programs should include new statistical methods, observational procedures, and data processing techniques. An important area is the development of techniques which, on a short-term basis, can assist prediction of long-term effects. Receiving-water goals should be set to serve as engineering guidelines that will protect the most sensitive biological resources within the area under consideration. At the same time, the conversion of biological liabilities to biological assets must be sought. For example, it is becoming apparent that heated discharges can be used to improve the productivity of oyster culture of finfish harvesting.

There are increasing examples of improved interaction between biologists and decision-makers in solving waste disposal problems. Such interaction, initiated early in each problem area, aids greatly in the performance of the functions of both groups, and in protecting the best interests of the public.

SPECIFIC PROBLEMS

Pesticides

Most use of pesticides occurs on land, but persistent chlorinated hydrocarbons, especially DDT and its derivatives, have penetrated the ocean and coastal waters of the world and are related to worldwide ecosystem degradation. Concentration of such materials increases toward the top of food chains, so that the carnivorous species contain highest concentrations and generally show the most drastic effects. Reproductive success of carnivorous fish and bird species within the marine environment can be substantially reduced. Filter-feeding species, such as oysters, can accumulate some of these compounds to levels many thousands of times the concentration in the environment. The commercial usefulness of some species has been reduced, and public confidence in the quality of marine organisms as food has suffered.

The effects of pesticides in the marine environment are not fully understood, but almost every investigation has revealed evidence of present or potential damage from DDT and related persistent pesticides. They reduce photosynthesis in phytoplankton, upset normal nerve function, reduce reproductive success, impair enzymatic activity, and cause other significant interferences with biological systems. There is no evidence as yet on the possible carcinogenic effects on marine organisms or on man. Although some of these effects are fatal to individual organisms, the more important biological effects are sublethal to the individual but of major consequences to the species. These effects and their causes are unusually difficult to detect until damage is widespread. The prediction of effects is complicated further by the low solubility of pesticides in water and their selective concentration in lipids, so that they partition into living organisms. Analyses of appropriate organisms high in the food chain and others, like mollusks, which are known to concentrate pesticides, are therefore usually more relevant and informative than are analyses of water.

Chemical materials combining the characteristics of mobility, chemical stability, low solubility in water, high solubility in lipids, and broad biological activity may be widely dispersed in the marine environment, affect many species, and be impossible to control, limit in distribution, or "manage," once they have been released in the environment. Their terrestrial uses may have inevitable deleterious effects on coastal and marine resources, thereby reducing the optimal utilization of those resources.

Public Health Risks

The term "public health risk" is used here to apply to impairment of human health, as it may be related to waste disposal in marine waters. Direct effects are those that may result from bodily contact with such waters, and indirect effects are those stemming from the ingestion of, or contact with, marine animals or plants affected by such waste disposal.

The program for measuring the impact of waste disposal on the marine biota is described elsewhere. Inevitably, factors considered in this program will include the effects of a wide spectrum of substances, including bacterial and viral pathogens, heavy metals, pesticides, organic-metallic compounds, and parasitic protozoa and worms. However, the ability of finfish and shellfish to accumulate such substances to a much higher concentration than in the ambient waters requires that the human risk involved in such food sources be determined. Moreover, the absorption even of traces of some substances may so affect the odor and taste of some shellfish and finfish as to make them unattractive for consumption, even when no pathogenicity is involved.

As to direct effects on man by marine waters used for waste disposal, existing data are very inconclusive. Epidemiological correlation with bathing in such waters has been sought and not yet found, even in the case of unrestricted bathing at beaches in England discernibly polluted with untreated raw sewage.

Public health regulatory agencies have used coliform concentration standards effectively to protect fresh water for potability, bathing, and fishing. In transferring such criteria to marine waters, it was assumed that coliform organisms were similarly a presumptive indicator of recent fecal pollution, that similar relationships observed in fresh water between total and fecal coliforms persisted in saline waters, and that similar die-away patterns for coliform in fresh water also took place in the marine environment. Abundant evidence has accumulated that casts doubt on all of these assumptions. Nevertheless, such standards have become entrenched both for acceptability of marine waters for body contact, and acceptability of marine shellfish for food. This may be questionable in both instances. As one example, such standards may endorse the consumption of shellfish with abnormal concentrations of metals or bacterial or viral pathogens, whose correlation with coliform indices is unknown. In another instance, they may impose a heavy economic or social burden by denying the use of beaches to the public on the basis of a tenuous assumption.

The specific need is to find some practical and reliable technique to measure the real health hazards. This might utilize a selected test organism, either bacterial or viral, but all practical alternatives should be considered. Until such new indices are developed, the coliform type of standards will have to be continued. Although present standards represent levels of attainability rather than direct public health significance, one major benefit from their use has been the compulsion of higher degrees of treatment.

Sludges and Solid Wastes

Sludge beds are most often derived from particulate matter formerly in suspension. Solid materials, either in deposits or suspension, may have similar biological effects. Therefore, these two categories of waste are herein treated together.

Suspended solids and subsequent sludge deposits may be introduced into estuarine and coastal marine environments in several ways, and their biological effects will vary according to the mode of introduction and origin of the solids. Suspended wastes may be more or less continually discharged into the environment from domestic or industrial outfalls, or waste solids may be separated from sewage and processed at the plant and intermittently discharged into the environment. The ecological effects of these two methods are known to differ but to what degree is not known.

A third method of discharge involves offshore barging of waste sludges. Whereas the other modes of discharge have a continual spatial if not temporal origin, the deposition of barged sludge varies with time, point of discharge, and prevailing current strength and direction. A fourth method of sludge formation involves the adherence of certain introduced organic and inorganic wastes onto "natural" particulates including both organic and inorganic particles. These aggregations then settle to the substratum where they accumulate as sludge beds which may be high in toxins.

Relatively little is known concerning the rate at which sludge deposits grow. Observations in the New York Bight have revealed, however, that barged sewage sludge has covered an area of several square kilometers during the past four decades. Moreover, it is known that uncontaminated sediments, and associated organisms, are carried onto the sludge beds during periods of severe wave action. Conversely, sludge deposits are frequently washed or carried over clean sediments. Sludge deposits measured in thousands of square meters are known to have accumulated around sewer outfalls. Adequate documentation does not exist in regard to the long-term stability of sludges or their rate of utilization by marine organisms.

Sludge accumulations have been observed to preclude completely normal marine life or to impinge upon the flora and fauna so as to reduce diversity and biomass. The effects on the biota may involve toxins contained in the sludge and suspended solids, direct burial and suffocation, anoxia resulting from reducing conditions, interference of suspended solids with normal feeding processes and reproduction, and change in community structure because of the competitive exclusion of one or more species by those species more tolerant of suspended materials or sludge deposits.

Sewage sludge deposits should be investigated to determine if they can serve to enrich local faunas, thus resulting in increasing the productivity of the water body. Since there is some evidence to suggest that sludge beds may be accompanied by the development of communities having low diversity and high biomass, these communities should be studied to determine if they are as productive in the food webs as are the natural communities unaffected by sludges and suspended solids.

Sludge beds are known to concentrate or contain toxic heavy metals, petrochemicals, and pesticides. The nature of these contaminants and their amounts depend upon the source(s) of the sludge and on those materials carried into the receiving waters from the land. Detailed analyses must be conducted to determine the presence of toxins in areas of sludge accumulation. Because of the variation in wastes, the composition of sludge deposits can be expected to vary both temporally and spatially. Consequently, their biological effects will vary. Differences in chemical content can serve as a tool to trace the movements and origins of sludge beds and suspended solids. Bacteriological examination similarly may be used.

In areas of high interest, bioassays should be conducted *in situ* and in the laboratory to determine the effects of known amounts of sludges and suspended solids on a wide variety of marine biota. Sludge fields and those water columns contaminated with waste-suspended solids must be monitored to determine the prevailing dissolved oxygen levels, reducing capacity, levels of toxicity, and other parameters of relevance. In the case of sludge beds, water samples should be obtained at the water-substratum interface. The number of samples collected and the frequency of collection will, as usual, depend upon the severity of the problem and the resources at hand.

A wide variety of fauna of diverse feeding habits should be investigated to determine the effects of sludge on suspension and deposit feeding taxa. The literature contains numerous references on this subject but very little is specific to sewage sludge and suspended waste materials. Recent work in England suggests that suspended sediments impinge adversely upon finfishes. Continual exposure to suspended material results in anomalous gill arch development. Such pathogenic effects should be investigated.

Temperature

Consumer demand for electrical power has resulted in unparalleled expansion within the electrical industry. Electrical generating capacities in several areas have been doubling every 7 to 10 years. To meet these demands, the industry has been locating more and more large-scale power installations in estuarine and coastal areas where a large portion of the demand exists, and where larger volumes of water for cooling uses are available.

New and expanded plants have enlarged to such an extent that they require large volumes of water (as large as 2,000 to 8,000 cubic feet per second) for the once-through condenser systems.

The environmental consequences of these activities have led to a considerable amount of research which is currently underway to ascertain the response of marine organisms to heated discharges in specific areas. Questions to which this research currently is seeking answers include:

1. What are the temporal and spatial distributions of all levels of marine organisms in response to existing thermal discharges in various latitudes and hydrographic conditions, and how are they affected by mixing and cooling phenomena?

2. What is the effect of the entrainment of planktonic and pelagic organisms within the cooling waters, and of their passage through the cooling system?

3. Under laboratory conditions, what are the responses of significant organisms to simulated thermal gradients and rates of change?

The research underway appears to be of relatively high quality. New needs for power in areas adjacent to previously unused coastal and estuarine waters of varying latitudes, temperature and hydrographic characteristics, will require the attention of aquatic biologists and other scientific and engineering specialists to insure the following:

1. That selection of new sites and expansion of existing plants is consistent with long-range water-resource objectives of local and regional plans.

2. That plant operations insure the preservation and promotion of aquatic life under conditions of variable water quality which may sometimes contribute to synergistic effects.

3. That design criteria for such facilities include adequate provisions for intake and discharge configurations to minimize the overall biological impact of the facilities.

4. That thermal power plants be operated in such a way as to mini-

mize the biological consequences of such daily activities as condenser cleaning and fouling control, as well as existing fluctuations in water quality.

It is important to recognize the need for input from specialists from many areas of marine science and engineering into decisions of location, design, and operation of facilities. Many serious biological problems can be avoided by prompt and proper consideration. Aside from acute and obvious ecological responses from thermal discharges, long-term operation of such facilities is believed to produce an almost infinite number of biological influences which, at the present time, are exceedingly difficult to assess quantitatively. Marine ecologists should make such evaluations under both field and laboratory conditions before decisions are made on the location of power-generating facilities in estuarine and coastal waters that are not well understood ecologically. Further, studies of the beneficial use of heated discharges for the promotion of aquaculture and agriculture should be investigated and thoroughly evaluated under differing conditions of latitude, water quality, and other significant variables. Such beneficial use of waste heat from power installations might be made to improve the estuarine and coastal fisheries, and should not be overlooked by the scientific and engineering professions.

Oil Spillage

Consideration is given here only to the biological implications of oil contamination and of methods used for its treatment. No attempt is made to discuss how oil may be dealt with, although the problems arising from treatment of oil by accepted techniques are described.

Oil on the surface of the sea does not usually cause extensive harm to marine life, although sea birds which dive through an oil-coated surface or come in contact with oil by other means usually die. Seals may also be affected. There is little evidence of acute toxicity on shore life, although sea urchins have died as a result of exposure to a light oil and the physical effects may be biologically destructive. Chronic exposure to oil and petrochemicals that can occur near major oil refineries usually causes a reduction of the biota over a wide area.

The ultimate fate of oil released into the marine environment is not fully understood, but bio-oxidation by marine micro-organisms is responsible for degradation of most of the oil which finds its way into the sea. There is a need to understand these mechanisms, and how they affect the long-term productivity of the sea. Oil and petrochemicals can cause tainting of finfish and shellfish with serious economic consequences; there is a need to understand this phenomenon, and to determine the concentrations and the types of substances which can cause tainting so that proper water quality criteria can be established.

Some of the biological problems associated with oil spillage result from the methods used to remove the oil. There should be additional research to develop efficient solvent-emulsifiers (detergents) which are nontoxic to marine species, especially in their developmental stages. There also is a need for efficient emulsifiers of lower toxicity than those currently available.

Removal of oil from the sea by the use of sinking agents requires further attention. Oil treated in this manner will affect the benthos, including sedentary commercial species, and these effects should be assessed. Sunken oil can foul trawls and other fishing gear which then have to be destroyed, and even small quantities of oil caught in a trawl can make the catch unacceptable. There is a need to understand in what areas and under what conditions sinking agents can be used without impairing fishing activities or affecting commercial shellfish. The fate of oil on the sea bed, its possible movement by currents, and its ultimate degradation require investigation.

In view of the very great difficulties in treating oil once it has become stranded on the beach or on rocks, further work is required to find methods of treating it while at sea—methods which are acceptable on biological grounds. Although sinking agents and emulsifiers may be used to minimize the effects of oil, they do not remove oil from the marine environment, and should be regarded only as a short-term method of dealing with the oil problem. It is strongly recommended that in the treatment of an oil spill to satisfy the requirements of the recreational uses of water and beaches, the overall biological effects inherent in the method of treatment should not be overlooked. Major effort at the local, national, and international level should be directed toward the prevention of oil spillage.

Toxic Substances

In addition to specific toxicants, such as pesticides and petroleum compounds which are discussed in other sections, numerous inorganic and organic ions, complexes and compounds require continual review to evaluate their impact on biological systems. Many of the metals, especially the transition elements, reach marine waters from an almost infinite variety of sources. Further, many of these elements are essential to marine organisms and are thus concentrated through specialized physiological mechanisms. A few examples follow:

1. Metals such as copper, zinc, chromium, nickel, lead and many others are discharged routinely. Bioconcentration by filter feeding and higher marine organisms facilitates the concentration of such elements at thousands of times the levels found in solution or suspension. Detailed knowledge of their effect upon various life-cycle stages of most estuarine and coastal species is lacking.

2. Radioisotopes of the transition elements discharged from atomic reactor waste treatment systems should be studied to evaluate their effect upon the marine biota. The studies should include, especially, the isotopes of elements such as cobalt (e.g., ^{58}Co and ^{60}Co), which are likely to be taken up as cobaltamin, a main constituent of an almost universal vitamin, B_{12}. The incorporation of tritium into nucleoproteins is not currently fully evaluated as to its potential influence on nutrition and other highly important cellular transformations. In a broad sense, these types of materials must be considered as important toxicants, and their incorporation into shellfish and finfish that are consumed either directly or indirectly by man must be viewed with concern.

3. The addition of wide-spectrum biocides for surface fouling control (including slime) is another example of the use of substances that might prove to be toxic to organisms residing in waste discharge areas. All such materials must be viewed with caution to prevent the use of either broadly effective, persistent, or chemically unpredictable materials for such purposes.

4. Toxic byproducts of the chemical formulation of commercial products such as pesticides may be discharged on either a continuous or batch basis into estuarine and coastal waters with little or no understanding of their effects upon the broad spectrum of desirable organisms resident in the discharge area. Even less is known of their influence on the critical life-cycle stages of both resident and seasonally transient species. It is recognized that monitoring such discharges is exceedingly difficult because of the lack of detailed knowledge of their chemical composition and environmental stability; however, their potential biological influence on other uses of surface waters must not be overlooked. Such materials should not be approved for discharge until their fate and effects are adequately understood and accepted.

Nutrients

"Biostimulation" is a general term usually applied to describe the complex set of factors involved in the growth of algae (and other organisms) in a receiving water due to the addition of nutrients (biostimulants). In some cases, the addition of large quantities of biostimulant produces excessive growth of algae and other organisms resulting in undesirable conditions. Sewage effluents contain considerable quantities of such nutrients except when tertiary treatment is applied to remove them. Large quantities of sewage from primary and secondary treatment systems now reach estuaries and coastal waters. Predicted population increases and the considerable cost of tertiary treatment both indicate that overenrichment may become a serious problem.

The short-term effects of excessive enrichment are generally rapid growth or blooms of algae, resulting in large fluctuations in diurnal oxygen concentrations, lowered dissolved oxygen due to algae die-off and biodegradation, and possible benthic animal and fish kills because of oxygen stress. According to present knowledge, such situations are more likely to occur in inland streams and lakes than in estuaries, and rarely, if ever, in the ocean. An attendant problem is the decrease in aesthetic quality of the receiving water, with the most serious problems occurring, again, in inland waters, and the least serious in the ocean.

Long-term effects include an increased rate of eutrophication, or aging, of the body of water. Characteristics of this process are increased plant production, shifts in the species composition, and a net increase in plant and animal biomass because of increased flow of food through the food chain.

One of the ways in which this process can be slowed is by limiting the addition of biostimulatory materials. Control of any single nutrient will be effective only if that nutrient can be reduced to a level low enough to be limiting to algal growth. These substances include not only the so-called macronutrients, such as nitrogen, phosphorus, and silica, but also the micronutrients which encompass a vast array of metals, vitamins, and other organic compounds which are needed (sometimes in minute quantities) for plant growth. Research in many parts of the world shows that waste-water effluents contain biostimulatory materials other than nitrogen and phosphorus. In other words, the nitrogen and/or phosphorus concentration may not account for all of the algal growth resulting from additions of waste water.

Since knowledge of these critical substances is limited, it is necessary to develop a quantitative measure of gross biostimulatory material.

The basic objective for conducting biostimulation tests is to obtain as accurate an estimate as possible of the near steady-state level of objectionable algae that will result from, or be supported by, a specific concentration of rate-limiting nutrient. These estimates represent the maximum concentration of the test organism (particular alga) that can be maintained in a given system that is at or near steady-state levels, and the maximum rate of growth that can be attributed to the concentration of the nutrient in the receiving water.

Ideally, results obtained in the laboratory should apply as accurately as possible to the specific practical problem and situation of concern. This is, or should be, the object of every bioassay procedure that is intended to provide a basis for establishing allowable concentration of contaminants in the environment. Probably the best example of this approach to "real" problems is the classic Biochemical Oxygen Demand (BOD) test and its use in the engineering design of waste treatment processes. This test is also used in approximate kinetic modeling of oxygen concentrations (oxygen sag curve) in streams as a function of time, organic loading (BOD), and numerous other physical and biological factors. The lack of precise simulation of the prototype by the laboratory assay method is well understood. However, the laboratory assay method in modeling the prototype does provide an approximation of the actual situation that is valuable in analyzing the problem and identifying practical design solutions.

It appears that such an approximation is exactly what is needed (a) for the crude but quantitative assessment of biostimulation or enrichment potential of all waste discharges, including both energy (carbon) and nutrient-bearing wastes, and (b) for prediction of allowable levels of organisms (algae) and nutrients of concern in the receiving water. Moreover, these data are susceptible to laboratory evaluation. The accuracy of such predictions (simulation of the prototype) can only be assessed by laboratory determination of the kinetic characteristics, followed by assessment of the resulting conditions in the field.

Methods for assaying biostimulation are under development. Both batch and continuous flow culture methods have been considered but it appears that only by continuous flow culturing can the prototype be reliably simulated.

Treatment of waste discharges to decrease their biostimulatory characteristics is not well understood. Recent research findings indicate that the solution to the enrichment problem may not be solved simply by the removal of 80 to 90 percent of nitrogen and phosphorus in waste discharges, such as is possible with tertiary treatment. The problem is much more complex, and is deserving of considerable research effort.

With reference to open coast waste disposal, the ideal objective is to control the addition and concentration of biostimulatory materials so that desirable and controlled enrichment of the coastal shelf is attained. Much needs to be done for us to understand, as well as to evaluate quantitatively, the desirable levels of fertilization or enrichment of the coastal waters through properly designed and controlled waste disposal practices.

Dissolved Organic Compounds

Two distinct types of dissolved organic substances are found in wastes—those which do not naturally occur in the marine environment, and those which are normally present to some degree.

Exotic compounds include some of known or suspected toxicity. Every effort should be made to assess their influence on marine organisms and to manage them properly. Naturally occurring compounds may be somewhat more difficult to study and monitor because of the problem of variable background concentrations. However, this category includes compounds of known biological activity (water-soluble vitamins for example). If concentrations of such compounds are introduced to populations that have had no previous "genetic experience," the results may be unpredictable. Dissolved organic compounds may play a significant role in the energetics of marine food chains.

It is important that information be obtained on the amounts of these compounds being discharged and their rates of dilution to permit comparison with background concentrations. Basic studies of dissolved and aggregated organic compounds, of their physiological effects, of their roles in food chain dynamics, and of their possible effects as chelating agents, should be expanded.

Oxygen Demand and Biological Degradation

A common characteristic of most wastes reaching coastal waters from terrigenous sources (tributary streams, domestic and industrial waste effluents either treated or untreated, storm run-off, and piped or barged sludges) is the presence of organic matter that undergoes biological degradation. This action can be detected by the increase in numbers of a few adaptable organisms as contrasted to clean waters that support fewer numbers of a large variety of species. In the process of stabilizing this putrescible fraction of the waste, oxygen is utilized. This biological action is an essential element in the assimilation of wastes, resulting in the maintenance of a balanced environment and thus assuring the use of the coastal waters as a renewable resource.

The utilization of oxygen from estuaries or restricted near-shore waters is one of the significant results of putrescible wastes, and one that may lead to serious biological consequences brought about by low dissolved oxygen or, in extreme cases, by its absence. In deeper water, dissolved oxygen depletion usually is not measurable; however, the possible long-term effect on restricted portions of the coastal waters should not be dismissed.

There have been many approaches to the measurement of putrescibility including both biological and chemical tests. The most maligned test, albeit the most acceptable as of now, is the Biochemical Oxygen Demand (BOD) test, referred to earlier. Fundamentally, it is a measurement of organic degradation under controlled conditions. It endeavors to reproduce in the laboratory the field conditions of a body of water that receives an organic load. In a true sense, it is a bioassay, a measure of biological activity. Also, it can reflect, in some instances, a measure of waste toxicity when inhibitors prevent or delay oxidation. The BOD test is the best available single overall measure of waste strength, and has withstood critical evaluation to become universally accepted and a part of waste-control language.

Past studies and interpretive data permit generally accepted measurement of the putrescibility or strength of wastes, especially by the BOD test. Its reproducibility is in the range of 5 to 10 percent. One weakness is the time needed to run the BOD test, since 5 to 30 days may be required. The test reflects biological stimulation and selectivity, as generally the carbonaceous material is first acted upon, to be followed in about 5 to 10 days by the development of nitrifying organisms. This phenomenon gives rise to a supplemental rapid oxygen demand known as the nitrogenous BOD. There is some evidence that occasionally nitrogenous demand initially can proceed concurrently with carbonaceous BOD.

Further study and research should be undertaken to understand more fully the dynamics of the BOD and related test, and to develop more sensitive methods. The following studies are suggested:

1. Development of a test that can be conducted over a short period of time that will provide information equal to or better than the BOD test.

2. Development of an alternate acceptable test that will be easier to conduct, and more susceptible to reproducibility.
3. Development of a BOD or alternate test that can be used in a remote monitoring program.
4. Study of the organisms involved in biochemical oxidation in the saline environment.
5. Study of the effects and measurement of nitrogenous BOD.

Brine Wastes

Continued research and interest in seawater conversion systems, as well as the need for alternate cooling systems such as cooling towers in coastal regions, emphasizes the need to understand the response of marine organisms to brine solutions as well as the general effects of gross salinity variations. Special emphasis must be given to the osmoregulatory problems of marine organisms within brine disposal zones, especially in relation to the effects of osmotic stress upon other environmental requirements such as temperature and oxygen needs. Research on such wastes is not extensive at present, and consideration of the impact of such local increases in salinity is essential.

Freshwater Discharges to Estuaries

Estuaries are normally defined as partially enclosed areas where waters of oceanic and terrestrial origins are mixed. One of the predominant features of an estuary is the salinity gradient from its mouth upstream. The location and slope of the gradient is determined by the amount of fresh water that enters the estuary, by tidal action, and by mixing. Geographical areas of an estuary may be designated as demonstrating a certain salinity range and stability, and the flora and fauna found in these areas reflect these limits.

Certain biological communities require rather stable salinity change patterns for survival, others must have periodic freshwater pulses, and some species and communities thrive best with wide temporal variation in salinity and other environmental conditions. Whatever biological community develops in any given pattern of salinity, the integrity of the community may be significantly altered locally by periodic fresh water inputs.

Freshwaters which enter estuaries may be classified as natural runoff and as waste effluents. In many instances, the waste effluents represent a sizable portion of the total freshwater inflow.

Natural runoff into an estuary may be considered to have "polluting" effects if the volume of the runoff fluctuates sufficiently, and with enough frequency, to cause severe osmotic stress on the biological system. It may also have deleterious effects due to waste materials entrained from overland flushing (such as floatable materials, detergents, and organic materials).

Waste effluents, depending on their volume and the location of the point of discharge, may significantly alter the ecology of an estuary on a short-term or long-term basis. This may be due either to the freshwater input or to waste materials carried by the effluent. Changes in the volume of effluent discharge, for example, from seasonal industrial discharges, storm discharges, or from combined sewer overflows, can affect the salinity gradient or cause localized decreases in salinity sufficient to impose severe osmotic stress on the biota.

There is urgent need for improved comprehension of the effects of alteration of freshwater inflow on the quality and uses of coastal waters. As industrial operations, river storage, and diversions increase, advance predictions of the principal effects of increasing or decreasing freshwater flow, or of modifying the seasonal flow patterns, will become essential.

Chapter 6
Recommended Research and Investigation for Effective Coastal Wastes Management

Effective rational management of the growing volume and variety of wastes generated by our accelerating coastal urbanization requires immediate initiation of a coordinated, long-term national program of research and investigation involving government, industry, and universities. When such a program has begun to supply answers to the many questions raised in the preceding chapters, we can begin to expand effectively our present waste treatment facilities commensurate with the task of maintaining and even enhancing the quality of our coastal waters.

Toward this end, we present here our recommendations for action as developed at the National Academy of Sciences–National Academy of Engineering Coastal Wastes Management Study Session. They are organized to reflect our assessment of the areas in which effective management of society's wastes is limited by lack of knowledge.

Our recommendations do not comprise an exhaustive catalog of information deficiencies in coastal marine science and engineering. Rather, they represent our assessment of a reasonable first step among the many programs of basic long-term research, the design-related investigations, and the collection of specific laboratory and field data, needed for improved design, management, and evaluation of coastal wastes treatment systems.

Our nation enters the present era of awakened and increasing public concern for effective wastes management with an existing and substantial framework of facilities, knowledge, organizations, and competent personnel in the area of coastal wastes management. This framework constitutes a formidable resource for maintaining and enhancing the quality of the environment, and provides a basis for the evolution of expanded and more effective mechanisms for applying scientific and engineering expertise to the problems in wastes management.

GENERAL RECOMMENDATIONS

Concept and Criteria for Waste Treatment

One of the greatest contributions that scientists, especially biologists, can make to conserving marine values is to furnish quantitative guidelines to assist the engineers who have responsibility for designing waste-treatment and disposal systems. The design of such systems must become much more scientifically oriented than in the past. Historically such design has been concerned primarily with maintaining aerobic conditions in the receiving waters and in keeping these waters safe for human health. *This criterion is no longer sufficient.* Methods are becoming available for assessing a broad range of marine receiving-water values. Thus the engineer's design should become less based on the use of "standard" systems and instead be tailored to preserve and enhance the specific receiving-water values of concern.

Professional Development and Institutional Arrangements

In response to the increasing emphasis on preserving and enhancing the quality of receiving waters, it is essential that the existing organizations and scientific and engineering expertise in coastal wastes management be used as a basis for the evolution of new and improved organizations and professional competence. Particular attention should be given to initiating or improving:

  1. Coordination of scientific research and engineering investigation, with emphasis on dissemination of the information generated.
  2. Planning for multiple use, including preservation, of the coastal waters and estuaries. Special attention should be given to the strength of local initiative in planning and operation within criteria reflecting local, regional, and national interests.

3. Development of regional policies, goals, criteria, and review procedures concerning resource development and use, including management of wastes, as they affect the coastal zone.

4. Allocation of a fraction of the cost of new wastes treatment systems and facilities to a program of monitoring waste discharges and receiving waters related to the facilities.

5. The quantity and quality of graduate education combining the interests of oceanography, ecology, and engineering.

6. Designation of research preserves to facilitate experimentation in estuaries and coastal waters in which the intrusion of other human influences is minimized.

7. International mechanisms for controlling persistent toxicants, such as chlorinated hydrocarbons, on both a worldwide and regional scale.

RECOMMENDATIONS CONCERNING MONITORING OF WASTE DISCHARGES AND RECEIVING WATERS

Research in Support of a Monitoring Program

Implementation of an effective monitoring program requires the initiation of specific research projects to improve the monitoring capability. Recommended research and development projects include:

1. Develop uniform sampling procedures for mass emission rates and receiving water, with consideration of the requirement for data processing.

2. Develop methods for quantitation of floatable matter and films and for identifying their origin.

3. Review the methods for detection of persistent pesticides.

4. Develop a method for quantitation and classification of persistent organics.

5. Develop a method for quantitation of gross heavy metals and gross acute toxicity.

6. Develop methods for monitoring biostimulants and for interpreting and applying the data.

7. Develop a method for quantitative description of biomass.

8. Develop methods for monitoring long-term effects in a community structure and its productivity.

9. Develop methods of monitoring trace metals (sediments).

10. Develop a method for quantitation of specific organics, especially in trace concentrations.

11. Identify the criteria necessary to define properly the wastes discharges that should be included in the category "significant" waste discharge.

A Monitoring Program

1. A program to monitor waste discharges and receiving waters should be initiated.

2. Characterization of wastes and receiving waters should take cognizance of the need for rapid, accurate, and economical methods for measurement of the selected parameters. In addition, instrumentation should be adapted or developed to perform the analyses and to transmit or record the observed data. Data analysis techniques should be developed so that corrective action can be initiated promptly.

3. Monitoring specifications must be examined periodically to insure their continuing adequacy and to remove redundancy.

Monitoring Waste Discharges

1. To implement a program of monitoring waste discharges, specifications should be developed for a core minimum program to be applied to all "significant" waste discharges. "Significant" waste discharges are to be defined as a result of a research project recommended above.*

2. The general objective of the core, waste-discharge monitoring program is to provide the minimum information needed to assess adequately the pollutional contribution of waste materials to the Nation's coastal environment. Specific objectives would include but not necessarily be limited to the following:

    a. Provide quantitative information on the unit and total mass emission rates for the common significant groups of wastes from significant waste-generating activities such as municipal, industrial, agricultural, natural, and other sources so that:

        (1) Adequate data are available for forecasting future waste contributions, based upon the level of future estimated waste-generating activity (population, industrial production, etc.);

        (2) Accurate input data are available for use in various model-

---

*The term "significant" is discussed in Chapter 2.

ing systems to provide estimates of waste concentrations and their variation in space and time; and

(3) It is possible to correlate or develop functional relationships between waste emission rates and waste effects that are principally biological in character.

b. Assess performance, on a gross basis, of waste treatment installations.

c. Insure that adequate information is available to permit improvements in waste treatment and disposal system design and operation.

3. All samples (except for "grab" samples collected for special analyses for high decay rate constituents) collected for routine analysis should be near-continuous, proportional, composite samples which accurately represent the characteristics of the waste stream (i.e., floatable, suspended, and dissolved constituents) with respect to their true mass emission rates (i.e., lb/day). Sufficient samples should be collected to provide an adequate statistical description for both the constituent concentration and the mass emission rate of the contaminant. After the waste has been statistically defined, analyses not pertinent to the local problem or to the wastes characterization should be deleted.

4. The analyses indicated in Table 10 should be conducted on essentially all samples collected.

5. Information on the accuracy and precision of both the sampling and analytical methods should be obtained and reported.

6. Data should be obtained on the level of waste-generating activity (i.e., for municipal waste—population tributary; for industrial wastes—tons of each product/day; etc.) so that waste discharges can be reported on a unit mass emission rate basis.

TABLE 10 Recommended Core Program Analyses—Waste Discharges

Floatable matter	Method needs development
Total and organic suspended solid	Methods adequate
Acute toxicity	Method needs review
Persistent pesticides	Method needs review
Persistent organic compounds	Method needs development
Biostimulants	Method needs development
Gross heavy metals	Method needs development
Coliforms (or equivalent)	Method under continuous review
Radioactivity	Methods adequate

Monitoring Receiving Water

1. For implementation of an effective program of monitoring receiving waters, the objectives of the program should include:

   a. Provide intermittent or continuous characterization of the receiving body of water and its terrestrial and atmospheric interfaces. Measurements sufficient to define the significant nature of the water body throughout a time period should be specified on the basis of statistical validity.

   b. Provide a knowledge of all sources of mass movement into and residence time within the receiving-water body, establish the significant character of such sources, and evaluate the relative contribution of each to the nature of the water body.

   c. Provide for rapid data evaluation and indicate the response procedures appropriate for the given water condition.

2. Monitoring program data should be obtained with consideration of the following factors:

   a. Sampling procedures which provide samples representative of the condition of the air, land, and water interfaces at any time.

   b. Sufficient vertical and horizontal control points, so that the samples will adequately describe the system.

   c. Sufficient frequency of sample collection to validate the analyses within any preselected statistical confidence limits.

   d. Analytical procedures that are of defined precision in terms of the parameter being measured.

The character of one restricted water body or coastal regime is quite likely different from another; therefore, no detailed recommendation can be made concerning the items b, c, and d above without enumerating the definitive characteristics of each water body. This analysis hopefully will be accomplished by a monitoring program with enough sampling locations and with sufficient frequency to describe the system within reasonable confidence limits.

Table 11 presents a summary listing of the recommended core program analyses of the waters and sediments. It outlines the recommended application of the tests to either restricted waters, the open ocean, or both for assessment of the condition of receiving waters and the effect thereon of the discharge of treated effluents. The core minimum monitoring program is not intended to be applied in its entirety to all marine waters but only to those bodies of water that receive "significant" waste discharges.

TABLE 11 Summary of Recommended Core Monitoring Program Analyses—Sediments and Water Column of Receiving Waters

Analyses	Applicable Region	
	Restricted Water	Open Ocean

SEDIMENTS

1. Physical
   a. Particle size distribution (methods adequate) — X / X
   b. Temperature (methods adequate) — X / X
   c. Other observations may also be needed for particle density, in-place density, and thickness of waste deposits to permit an estimate of the volume and mass of wastes accumulated (techniques need evaluation) — X / X

2. Biological
   a. Quantitative description of the standing crop of benthic organisms (quantitative technique needs development) — X / X
   b. Other tests including an index of bottom respiration may be useful to indicate the amount of readily biodegradable organic matter in the deposit (technique needs development) — X / X

3. Chemical
   a. Concentration of organic matter by concentration of organic carbon or organic nitrogen (technique needs evaluation) — X / X
   b. Presence or absence of H_2S (quantitative technique needs evaluation) — X / X
   c. pH (technique adequate) — X / X
   d. Other measurements should be made for suspected toxicants when appropriate including specific trace metals (technique needs evaluation) — X / X

TABLE 11 (Continued)

Analyses	Applicable Region	
	Restricted Water	Open Ocean
WATER COLUMN		
1. Physical		
a. Quantification of floatable material and films with analysis for determination of probable origin of material (method requires development)	X	X
b. Water clarity by photometric or other methods (methods adequate)	X	X
c. Temperature—continuous recording with depth or at least three points in vertical column (method adequate)	X	X
2. Biological		
a. Coliform determination (method needs evaluation)	X	X
b. Biostimulatory characteristics (method to be developed)	X	X
c. Assessment of biomass including standing stock and community structure to determine long-term effects of waste discharges (techniques to be developed)	X	X
3. Chemical		
a. Dissolved oxygen (method adequate)	X	
b. Chlorosity (method adequate)	X	X
c. pH (method adequate)	X	X
d. Nitrates (method needs periodic evaluation)	X	
e. Phosphates (method needs periodic evaluation)	X	

RECOMMENDATIONS CONCERNING PHYSICAL PROCESSES AND INTERACTIONS

Initial Dilution and Diffuser Design

1. Present knowledge of buoyant jet diffusion is nearly adequate for design of outfalls (including a multiple port diffuser) to achieve a pre-

scribed initial jet dilution and submergence below any given thermocline. However, further research is needed in a number of areas. Primarily, there is need for understanding of line sources, and how well multiple-jet diffusers may be represented by line sources. Although current effects on initial plume behavior are not well understood, they are not as critical a factor as density stratification in predicting initial dilutions due to jet mixing.

2. Methods do not exist for predicting the size and shape of the waste fields (of either conventional or heated effluents) that are developed at the end of the initial jet-mixing stage. Closely coupled with this shortcoming is the problem of lateral spreading due to density differences between the field and its environment. Research should be conducted on both of these problems.

3. For barge dumping of sludges in the ocean, research is needed on flows generated by suddenly released sinking sludge in a stratified environment.

4. Control of thermal waste in coastal waters involves the same kind of stratified flow problems as sewage disposal. Inasmuch as large submerged diffusion structures are not yet in use, some problems of large single jets need special study, such as the behavior of a buoyant surface jet injected in a stream perpendicular to the current.

5. Field studies of flow patterns and dilutions over waste outfalls are needed urgently to confirm design predictions and methods. Most of the hydrodynamics of buoyant jet mixing has been confirmed only in laboratory experiments. Similarly, the effects of hydrodynamic forces on the diffuser structures themselves require continuing investigation.

Physical Processes in Estuaries

1. It is necessary to develop a sound physical basis for quantitative predictive models of time and space variations of constituent distributions in estuaries. This project will require further work on theoretical, numerical, and physical models, including determination of the correlation between the models and field studies.

2. Further knowledge is required of the relationship of the mean circulation, tidal currents, and turbulent exchanges to the river inputs, external tides, external density distribution, wind, and the shape and size of the estuary.

3. Little is known of conditions responsible for the change in an estuary from a salt-wedge to a partially mixed estuary, or from a fiord to either a salt-wedge or a partially mixed estuary. These conditions need study, particularly in fiords.

4. In the development of models, both theoretical and numerical models should be stressed, as they include the possibility of the incorporating of biological, chemical, and physical processes at prototype scales.

5. Turbulence processes need investigation, as their dependence on density stratification and mean-velocity shear plays a dominant role in the behavior of estuaries.

Turbulent Flux and Diffusion

1. Detailed observational approaches to the problem of turbulent diffusion are needed. Simultaneous measurements of turbulent fluctuations in velocity, salinity, and other properties, together with environmental factors such as shears in mean velocity and stability of the water column are necessary. Likewise, tracer studies on a scale of 10–100 meters should be carried out under various environmental conditions.

2. There is need to develop predictive models for gross spreading of patches and plumes in the ocean from the combined effects of eddy diffusion (both horizontal and vertical) and shear in the mean velocity field. Item 1 above recommends steps that will provide a basis for this development, and will allow a better interpretation of previously reported values of gross dispersion coefficients.

3. Systematic tracer experiments should be carried out in subsurface waters in order to have more reliable information on the dispersion of patches or plumes. These experiments should include the use of artificial tracers, such as fluorescent dye, and studies of existing waste fields which occur at subsurface depths.

Physical Processes in Coastal Areas

1. For a proper understanding of coastal circulation on all scales, a program of collection of oceanographic and meteorological data is recommended. The observations should be made over a long enough period of time to reveal all periodicities up to and including annual. Although such a program could be carried out by multiship operations, moored arrays of instruments capable of sampling the entire water column would probably be better. Such a program should permit evaluation of wind, river inflow, tide, and internal waves as transport mechanisms.

2. To improve our ability to predict the fate of wastes introduced into estuaries and coastal waters under specific environmental condi-

tions, a study is recommended of the effects of intermediate scale variations in the current pattern on the time-varying concentrations of waste components at various distances from the source, using tracers such as fluorescent dyes as well as waste components from existing outfalls.

3. The large-scale processes which lead to exchange of coastal water with oceanic water should be studied. Development of a fluorometer capable of sampling at all depths, which is an order of magnitude more sensitive than any available at present, is needed so that large-scale dye-tracer experiments can be carried out economically. Alternatively, a more economical tracer might be developed for such work.

Decay of Nonconservative Constituents as Related to Physical Factors

1. A series of controlled field experiments should be conducted to study the nonconservative properties of such constituents of wastewater as enteric bacteria and other toxic substances discharged into coastal and estuarine waters.

2. As soon as reliable detection and enumeration techniques have been developed, these studies should be expanded to include pathogenic viruses.

Interactions between Floatable and Settleable Components of Wastes and Physical Factors

1. Studies should be conducted to ascertain the prevalence, properties, and character of floatables that originate from waste water and sludge (including barged materials) in coastal waters and in estuaries. The substances comprising the various forms of the floatables (particulate matter, films, scum, and foam) should be identified as to primary source.

2. Investigations should be made to determine the means by which the floatables are collected and compressed into slicks or streaks on the water surface, as well as the natural mechanisms available for transporting the materials in the water surface.

3. Studies should be made to ascertain methods of treating or handling the waste waters and sludges that will reduce or eliminate problems of surface pollution.

4. Studies should be conducted to evaluate the movement and dispersion of releases of sludge at sites currently in use, such as in the New York Bight and off southern California. These studies should in-

clude, but not necessarily be limited to, investigation of the methods of introducing the sludge, i.e., by barge or outfall, and the transport mechanisms, including settling and resuspension, which influence the distribution and spread of the materials.

RECOMMENDATIONS CONCERNING CHEMICAL FACTORS

Chemical Processes Involving Dissolved Inorganic Constituents

1. The concentrations and forms of trace elements believed to be biologically significant in the waters and sediments and their concentrations in organisms in different areas should be determined. Areas that should be examined are near the mouths of large rivers and coastal areas where freshwater inputs come primarily from waste-water discharges. The elements of concern would probably include but not be limited to copper, zinc, cobalt, chromium, arsenic, molybdenum, selenium, mercury, cadmium, and lead.

2. The degree of complexing of trace metals by the organic and inorganic constituents of waste-water effluents, sea water, and estuarine waters should be evaluated in both laboratory and field studies. Temperature ranges in the natural environment, as well as in the vicinity of thermal outfalls, should be represented in the experimental program. Not only may the degree of complexing prove significant in controlling the behavior of the metal ions, it may also prove pertinent to an understanding of the action of organic residues. The forms in which the metals exist are important factors affecting their biological activity.

Chemistry of Particles and Processes in Sediments

1. Experiments should be carried out to establish the effects on soluble components, particularly waste solutes, of flocculation, aggregation, coprecipitation, and sorption. A study should be made of the physical–chemical factors and the role of organisms in affecting the flocculation rates of sediments in estuaries and coastal waters. Pertinent variables appear to be the degree of dilution of fresh water suspensions entering sea water, the levels of organic matter, the pH of the mixture, the oxidation potential, the relative percentages of different clay minerals and other solid phases, the mixing characteristics of the flow, and the temperature.

2. The rates of aggregation and sedimentation of organic particles in

the marine environment should be studied. Such factors as pH, temperature, organic-metal ion complexing at organic particle surfaces, and the concentration of inorganic particles should be evaluated. Organic debris appears to play a role in transporting trace metals to the sediments. The organic debris may associate with inorganic particles, thus affecting the sedimentation of inorganic phases (oxides, clays, silica).

3. The biological and chemical transformations occurring in contaminated and uncontaminated sediments should be determined, with particular reference to nutrients and trace elements. These studies should include considerations of concentration gradients, movement of water at the sediment interface, eddy diffusion, and the release of gas on the rates of transport from sediments to the water column. Also included should be the effects of changes from oxidizing to reducing conditions, and vice versa.

4. Adequate procedures should be developed for distinguishing among inorganic particles, living organisms, and dead organic matter, both in the water column and in the sediments.

Nutrient Chemistry and Biochemical Changes

1. The fluxes of nitrogen and phosphorus in all phases of the cycles affecting the marine environment should be explored. The study should not overlook the fluxes due to rooted benthic plants, birds, and humans.

2. An understanding should be developed of the amount and character of dissolved and particulate organic matter in the ocean; its origin, including the contributions from rivers and waste discharges; its spatial distribution; and its biological significance.

3. A study of the factors that control the qualitative and quantitative aspects of phytoplankton blooms in estuarine and coastal waters should be carried out.

4. The effects of adding nutrients (phosphate, nitrate, silicate) and oxidizable carbon on the primary productivity and on the resulting organic load in restricted coastal environments should be determined. The relative effects of the individual nutrients are important considerations. The rates of oxygen exchange between the atmosphere and other sources (e.g., ferric oxide in sediments) and the coastal waters should also be studied. These studies will help provide a basis for predicting to what extent re-aeration can compensate for the oxygen demand caused by the introduction of oxidizable carbon and nutrients from waste outfalls. Factors such as wind stress, depth, pressure head, den-

sity gradient and stability, and surface films such as petroleum should be considered.

5. The biochemical mechanisms for concentrating trace components by the biota, the subsequent effects of this concentration on the organisms involved, and the transport and further concentrating of these trace components as they move up the food chain should be determined.

6. Subtle, sublethal effects of waste products on physiological and biochemical processes—such as enzyme induction or inhibition, ion transfer across membranes, and chemosensitive reception—should be studied. Effects of these kinds may significantly influence the growth, reproduction, development, or survival of marine animals in ways not detected by conventional assay or toxicity tests, or population studies. It is in this area of sublethal effects that ocean disposal of wastes may encounter its most serious problems.

The Chemistry of Specific Wastes

1. Even with the establishment of improved safety criteria and redundant emergency systems, the probability of the occurrence of oil leakage and bilge washings from ships, of catastrophic events such as shipwrecks, and of oil seepage and operating well casualties on the continental shelf, indicates that research is needed on:

    a. Natural biochemical processes responsible for degradation of oil films or oil droplets.

    b. Techniques of analysis for detecting and characterizing low concentrations of oil in water and for identifying sources.

    c. The effects of different oil dispersants in degradation of the oil, the toxicity of dispersant and dispersant-oil mixtures to marine organisms, and the uptake of the oil, dispersant, and/or dispersant-oil mixtures in the food chain.

    d. The effects of added settling agents on bottom characteristics and on the benthos, and the fate of oil so deposited.

    e. Fractionation of oil films on exposure to environmental influences, and the fate of residual materials in the sea.

    f. The effect of oil films on the air–sea oxygen exchange, and interference in processes of biological productivity, such as changes in light penetration and mixing.

2. The fluxes of synthetic organic chemicals into the ocean through sewage outfalls, rivers, atmosphere, and biota should be determined. Priorities should be given to potentially hazardous or deleterious materials such as pesticides, detergents, fuel residues, certain solvents, etc.

Chemical Consequences of Man's Physical Activities

1. The effects of human activities (such as forestry, agriculture, terrestrial and marine mining, dredging, and impoundments), on the flow of inorganic suspended matter to the oceans, and on the distribution and character of the sediments should be determined. Among the potentially significant effects are those on transparency of overlying waters, oxygen demand from reducing sediments, transport or release of nutrients including trace elements, alterations of the benthos, silting of harbors, and erosion of beaches.

RECOMMENDATIONS CONCERNING BIOLOGICAL EFFECTS

1. Studies should be made immediately of selected existing outfalls and disposal areas in several distinct marine biogeographic provinces. These studies, and the relationships derived from them, must serve as an interim basis for improved evaluation of the acceptability of new disposal facilities and sites. Completely adequate techniques are not available for definitive assessment of all impacts of wastes on coastal waters. The studies should include at least the following:

   a. Quantitative floral and faunal surveys in the immediate vicinity of discharge, within the measurable zones of influence, and at reference sites.

   b. Sludge fields (when present).

   (1) Measurement of the temporal and spatial dimensions of sludge fields.

   (2) Chemical analyses of sample sludges from various outfalls with emphasis on substances likely to have biological importance.

   (3) Measurement of the rates of biodegradation and utilization of sludge components by marine organisms.

   c. Determination of the dissolved inorganic and organic substances resulting from coastal discharges and their effects by:

   (1) A chemical inventory of components.

   (2) Bioassays of both effluents and affected waters for toxicity and stimulation.

   (3) A study of primary productivity and other community responses in affected waters.

2. A detailed examination of the public health significance of coastal discharges should be made, including:

   a. Re-evaluation of the adequacy of traditional fresh-water biological indexes in marine waters and in organisms consumed by man.

    b. Development and application of improved indexes.

    3. Research on the biological concentration of waste components by marine organisms should be expanded and intensified. Special attention must be given to organisms involved either directly or indirectly in the food chain of man, without sacrificing adequate attention to the complete environment.

    4. The input of DDT into the marine environment by the United States should be eliminated. To avoid repetition of the DDT type of problem, we further recommend that any material that combines the properties of mobility, chemical stability, low solubility in water, and high solubility in lipids be kept out of the marine environment unless it has been proven not to have the broad biological activity that is characteristic of DDT.

    5. Long-range, properly designed, detailed, quantitative studies of the structure and dynamics of animal and plant communities and their relationship to waste disposal in carefully selected areas should be established and supported. These areas should include those that are relatively little affected, that are being affected at an increasing rate, and that are already seriously affected. Some of the studies should be done in designated and protected marine preserves. All should be related to the beneficial uses to which the particular coastal region is allocated.

    6. Programs of physiological studies to define the tolerable limits of waste concentration for each of the specific uses envisioned for the coastal regions designated in a long-range plan should be established and supported.

    7. Programs of systems analysis and model development that will improve prediction of the biological effects of various possible combinations of waste treatments, disposal systems, and uses of the receiving water should be instituted and supported. As more data become available from the studies suggested above, models can be continually refined.

    8. All proposals for new installations, modifications, or activities that may result in major changes in the amounts or nature of the wastes should be reviewed to determine whether quantitative ecological studies of the biota are required, both before and after the change. If such studies would lead to greater protection of the biota, or would provide better bases for regulation, adequate funds for them should be included in the budget. Enough time must be allowed for careful studies, especially those to be done before the change is made. Data

from such studies would increase the accuracy of models and strengthen the objective bases for setting standards.

9. The U.S. Government should consider, and act effectively upon, the ultimate disposal problems and the biological effects of new products of any kind which, after release in the commercial market, could result in the impairment of the biological values of the marine environment.

Chapter 7
Suggested Priorities and Estimated Minimum Effort Required

Our recommendations select, from among the broad scope of scientific and engineering research and investigation program areas in wastes management, those projects that we believe are essential and that should be assigned high priority to improve effectively our wastes management practices.

We have assigned relative priorities to each of the recommended projects within each of the major program areas. The minimum effort required for effective results and the period required for completion of specific projects has been estimated for each project. *These suggested priorities and allocations of effort are, of course, highly subjective.*

Although priorities were estimated within each of the major program areas, no attempt was made to compare priorities in each of the areas with those in the others. On the other hand, the minimum effort that is suggested for each of the program areas compared with the others indicates our estimate of the relative emphasis to be placed in each area of the total initial minimum program.

Further detailed refinement of priorities should be undertaken on a continuing basis by those within industry, government, and universities who, because of their responsibilities and competence in developing and utilizing the results of the research and investigation, will be involved in operational and research problems. Continued refinement of the estimates of effort required beyond the suggested initial minimum

Suggested Priorities 101

effort, and the refinement of time required for the initial and any additional effort, should also be undertaken.

PROGRAM AREA OF MONITORING WASTE DISCHARGES AND RECEIVING WATERS

The recommended routine-type monitoring program should be initiated immediately, should be expanded to meet management information requirements, and should be improved as monitoring techniques resulting from the recommended specific research projects become available. The monitoring program should be a continuing and a regular part of the waste disposal operation. No estimate of required effort for the actual field-scale monitoring program is given.

Table 12 lists the relative priorities and estimated minimum effort for specific research projects that will be required to implement the broad recommendations for a program of monitoring and investigation of waste discharges and receiving waters.

For waste streams like those in agricultural and industrial areas, additional research and development on specific sampling and analytical methods is required. For receiving-water monitoring, there also will be special development efforts associated with particular monitoring problems. The magnitude of this research and development may be equal to, or greater than, that required for the core monitoring programs recommended in this study.

PROGRAM AREA OF PHYSICAL PROCESSES AND INTERACTIONS

Relative priorities and estimated minimum effort for the recommended research and investigation in physical processes and interactions are presented in Table 13. The relative priorities for each project have been estimated within each of six sets of related projects. The estimates of effort represent that which we believe is required to conduct the recommended programs at a level that will provide beneficial results.

PROGRAM AREA OF CHEMICAL FACTORS

Relative priorities and estimated minimum effort for recommended research and investigation in chemical factors are summarized in Table 14.

TABLE 12 Priorities and Estimated Initial Minimum Effort for Research and Investigation Needed for Improving Waste Discharge and Receiving-Water Monitoring Programs[a]

Research Required To Implement the Monitoring Program	Research Concerned with		Estimated Minimum Total Effort[b] (man-years)	Priority	Completion Time
	Waste Discharge	Receiving Waters			
Uniform sampling procedures					
Relative to mass emission rates, receiving waters, data processing	X	X	11	A	S[c]
Floatable matter					
Method of quantitation	X	X	11	A	S
Films					
Method of quantitation		X	11	A	S
Persistent pesticides					
Review method of determination	X	X	11	B	S
Persistent organics					
Method of determination, quantitation	X	X	13	B	S

Gross heavy metals				
Method of quantitation	X	7	B	S
Gross acute toxicity				
Method of quantitation	X	7	A	S
Biostimulants				
Methods and interpretation	X			
Biomass[e]		34	A	L[d]
Method and quantitative description	X	27	A	L
Community structure — productivity[e]		50	B	L
Methods for long-term effects	X			
Trace metals (sediments)[e]				
Method	X	11	C	S
Specific organics[e]				
Method of quantitation — trace concentration	X	13	C	S
Significant discharge				
Definition of	X	4	A	S

[a]The recommended Monitoring Program itself is not included.
[b]Total effort for this program area is 210 man-years.
[c]S is short-term (less than 5 years).
[d]L is long-term (less than 10 years).
[e]These projects must be examined in detail for compatibility with projects recommended under chemical factors and biological effects.

TABLE 13 Priorities and Estimated Initial Minimum Effort for Research and Investigation in Physical Processes and Interactions

Recommended Research and Investigation	Estimated Minimum Total Effort[a] (man-years)	Priority	Completion Time
Initial dilution and diffuser design	37		
Buoyant jet diffusion		B	S[b]
Waste fields		B	S
Barge dumping of sludge		A	S
Thermal waste		B	S
Flow patterns		A	L[c]
Physical processes in estuaries	185		
Quantitative predictive models		A	L
Hydrodynamics		B	L
Estuary transitions		A	S
Biological and chemical processes		A	L
Turbulence processes		A	S
Turbulent (eddy) flux studies	72		
Observational studies		A	S
Predictive models		A	S
Subsurface tracer experiments		B	S
Physical processes in coastal areas	360		
Data collection		A	L
Intermediate-scale current patterns		A	S
Large-scale exchange processes		B	S
Decay of nonconservative constituents as related to physical factors	20	A	S
Interactions between floatable and settleable components of wastes and physical factors	46		
Character of floatables		A	S
Mechanisms of transport		B	S
Reduction of surface concentration		B	S
Case studies		A	L

[a]Total effort for this program area is 720 man-years.
[b]S is short term (less than 5 years).
[c]L is long term (less than 10 years).

TABLE 14 Priorities and Estimated Initial Minimum Effort for Research and Investigation Needed in Chemical Factors

Recommended Areas of Research and Investigation	Estimated Minimum Total Effort[a] (man-years)	Priority	Completion Time
Trace metals	50	A	S[b]
Complexing	22	B	S
Inorganic aggregation	22	B	S
Organic aggregation	17	B	S
Diagenesis	13	B	L[c]
Distinguish organic vs. inorganic	5	C	S
Nutrient fluxes	22	C	S
Organic matter distribution	13	B	L
Phytoplankton blooms	42	A	S
Anoxic conditions	17	B	S
Biochemical concentration	17	B	L
Sublethal effects	34	A	L
Oil spillage	134	A	S
Synthetic organics	17	A	L
Human physical activities	25	C	L

[a] Total effort for this program area is 450 man-years.
[b] S is short term (less than 5 years).
[c] L is long term (less than 10 years).

These recommendations, listed as specific projects, are indicative of broad areas of investigation, within which re-emphasis may be desirable in the future.

PROGRAM AREA OF BIOLOGICAL EFFECTS

Priorities and estimated minimum effort for project areas of research and investigation on biological effects are summarized in Table 15.

TABLE 15 Priorities and Estimated Initial Minimum Effort for Research and Investigation in Biological Effects

Recommended Areas of Research and Investigation	Estimated Minimum Total Effort[a] (man-years)	Priority	Completion Time
1. Intensive study of outfall areas and effects	620	A	L[b]
2. Public health significance of wastes	25	B	S[c]
3. Study of biological concentration mechanisms	40	B	S
4. Management of DDT	–	B	S
5. The structure and dynamics of coastal biological communities	370	A	L
6. Defining tolerable limits for each major use	190	A	S
7. Improvement of systems and models	35	B	S
8. Criteria for review of proposals for ecological study requirements	–	A	L
9. Evaluation of new waste products	–	B	L

[a] Total effort for this program area is 1,280 man-years.
[b] L is long term (less than 10 years).
[c] S is short term (less than 5 years).

Appendix A

Steering Committee on Coastal Wastes Management

Donald W. Pritchard
Co-chairman
L. Eugene Cronin
John H. Ryther
Richard C. Vetter
Executive Secretary,
NASCO

Erman A. Pearson
Co-chairman
John D. Parkhurst
Richard D. Pomeroy
Russell Keim
Executive Secretary,
NAECOE

U.S. GOVERNMENT LIAISON REPRESENTATIVES

William A. Cawley

David G. Stephan

Appendix B

Participants in Coastal Wastes Management Study Session

7–12 July 1969, Jackson Hole, Wyoming

NASCO DESIGNEES
Harry H. Carter
L. Eugene Cronin
Edward W. Fager
Edward D. Goldberg
M. Grant Gross
Donald W. Hood
John A. McGowan
Akira Okubo
John B. Pearce
Donald W. Pritchard
Ricardo M. Pytkowicz
Maurice Rattray, Jr.
Francis A. Richards
John H. Ryther
Howard L. Sanders
Richard C. Vetter
Michael Waldichuk
Charles F. Wurster

NAECOE DESIGNEES
Neal E. Armstrong
Robert D. Bargman
Norman H. Brooks
William B. Davis
James E. Foxworthy
Norman B. Hume
Loren D. Jensen
Russell Keim
Martin Lang
Jan J. Leendertse
Harvey F. Ludwig
James J. Morgan
John D. Parkhurst
Erman A. Pearson
Richard D. Pomeroy
Ralph Porges
Robert E. Selleck
Jerome E. Stein

U.S. GOVERNMENT OBSERVERS

David A. Adams
J. Frances Allen
Donald J. Baumgartner
Harold Berkson
Richard Callaway
William A. Cawley
Norman Cutshall
Milton H. Feldman
Charles G. Gunnerson
John R. Hyland
William Lehr

H. William Newman
Kenneth Osborn
Donald K. Phelps
Ralph L. Rhodes
R. Lawrence Swanson
Robert L. Swart, Jr.
Clarence M. Tarzwell
Richard A. Wade

OBSERVER FROM UNITED KINGDOM

Peter C. Wood

Appendix C

Committee on Oceanography and Committee on Ocean Engineering

Committee on Oceanography
NATIONAL ACADEMY
OF SCIENCES

John C. Calhoun, *Chairman*
Richard G. Bader
Karl Banse
Wayne V. Burt
Charles L. Drake
Paul M. Fye
Howard R. Gould
John A. Knauss
Gerald J. Paulik
Donald W. Pritchard
Henry Stommel
George P. Woollard
Warren S. Wooster
Richard C. Vetter
   Executive Secretary

Committee on Ocean Engineering
NATIONAL ACADEMY
OF ENGINEERING

Thomas C. Kavanagh, *Chairman*
Walter C. Bachman
Leo L. Beranek
Antoine M. Gaudin
LeVan Griffis
James M. Hait
Edward H. Heinemann
Alfred A. H. Keil
John R. Kiely
Edwin A. Link
Arthur E. Maxwell
George C. Nickum
Erman A. Pearson
William E. Shoupp
Elmer P. Wheaton
Russell Keim
   Executive Secretary

Appendix D

Background Papers[*]

Monitoring

Bargman, R. D., and J. D. Parkhurst, Evaluation of Sources and Characteristics of Waste Discharges.
Davis, W. B., Monitoring of Marine Environment for Waste Components and for Effects of the Introduced Wastes.
Gross, M. G., New York City–A Major Source of Marine Sediment.
Pearson, E. A., Waste Discharge Monitoring.

Physical Effects

Carter, H. H., Physical Processes in Coastal Areas.
Foxworthy, J. E., Dispersion of Non-Conservative Wastes Discharge into the Ocean.
Okubo, A., Physical Processes Which Interact with and Influence the Distribution of Wastes Introduced into the Marine Environment.
Selleck, R. E., Physical Effects on Receiving Waters.

Chemical Effects

Hood, D. W., Chemical and Geochemical Effects on Receiving Water.
Morgan, J., and R. D. Pomeroy, Chemical and Geochemical Processes Which Interact with and Influence the Distribution of Wastes Introduced into the Marine Environment, and Chemical and Geochemical Effects on the Receiving Waters.

[*]Prepared by invited experts as a basis for discussion and development of recommendations by the participants at the study session. These are available from NASCO and NAECOE as a supplement to this report.

Richards, F. A., Some Chemical and Geophysical Processes Which Interact with and Influence the Distribution of Wastes Introduced into the Marine Environment.

Biological Effects

Armstrong, N. E., and P. N. Storrs, Biological Effects of Waste Discharges on Coastal Receiving Waters.
Cronin, L. E., Biological Effects on Receiving Waters.
Jensen, L. D., Biological Processes Which Interact with and Influence the Distribution of Wastes Introduced into the Marine Environment.
McGowan, J. A., The Coastal Pollution Problem in California.
Strickland, J. D. H., Biological Effects on Receiving Waters: The Plankton.

Bibliography

MONITORING

Alyakrinskaya, I. O. Experimental data on oxygen consumption in sea water polluted by petroleum. Okeanologiya 6(1):89, 1966.

Anderson, J. B., E. B. Henderson, and C. F. Weber. The role of benthos and plankton studies in a water pollution surveillance program. Proc. 18th Ind. Waste Conf., Purdue Univ., 275, 1965.

Azad, H. S., and D. L. King. Evaluating the effect of industrial wastes on lagoon biota. Proc. 18th Ind. Waste Conf., Purdue Univ., 410, 1965.

Bailey, T. E. Fluorescent-tracer studies of estuary. J. Water Pollution Contr. Fed. 38:12, 1968.

Baker, R. A. Characterization of the microorganic constituents of water by instrumental procedures. Proc. 22nd Ind. Waste Conf., 252, 1967.

Ballinger, D. G. Automated water quality monitoring. Environ. Sci. Technol. 2(8):606, 1968.

Balter, E., K. K. Turekian, and D. F. Schultz. The distribution of rubidium, caesium, and barium in the oceans. Geochem. Cosmochem. Acta 28:1459, 1964.

Bonde, G. J. Pollution of a marine environment. J. Water Pollution Contr. Fed. 39(10):44, 1967.

Burt, W. V., and G. F. Beardsley. Underwater optical measurements. Oceanography May/June: 35, 1969.

Chan, K. M., and J. P. Riley. The automatic determination of phosphate in sea water. Deep Sea Res. 13:467, 1966.

Chow, T. J. Isotope analysis of seawater by mass spectrometry. J. Water Pollution Contr. Fed. 40(3):399, 1968.

Cooper, L. H. N. Chemistry of the Sea II. Chem. of Brit. 1(4):150, 1965.

Copeland, B. J. Biological and physiological basis of indicator communities. In Pollution and Marine Ecology. Interscience Publishers, New York, 1967.

Curl, H., Jr., and E. W. Davey. Improved calibration and sample-injection systems for nondestructive analysis of permanent gases, total CO_2, and dissolved organic carbon in water. Limnol. Oceanog. 12:545–548, 1967.
Curl, H., Jr., and G. C. McLeod. The physiological ecology of a marine diatom, *Skeletonema costatum* (Grev) Cleve. J. Mar. Res. 19:70, 1961.
Curry, L. L. Midge larvae as indicators of radioactive pollution. Proc. 13th Ind. Wastes Conf., Purdue Univ., 269, 1960.
Danilov, F. J. The decade ahead. Oceanology May/June: 30, 1969.
Deacon, G. E. R. Chemistry of the sea I. Inorganic. Chem. of Brit. 1(2):48, 1965.
Duedall, T. W. Partial molal volumes of 16 salts in sea water. Environ. Sci. Technol. 706, 1968.
Edmonson, W. T., and Y. H. Edmonson. Measurements of production in fertilized salt water. J. Mar. Res. 6:228, 1947.
Edwards, G. P., A. H. Malof, and R. W. Schneeman. Determination of orthophosphate in fresh and saline waters. J. Amer. Water Works Ass. 57:917, 1965.
Ferguson, W. Factors to be considered in the abatement of sea and estuarine pollution. J. Proc. Inst. Sewage Purif. Part 4, 372, 1964.
Finger, J. H., and T. A. Wastler. Organic carbon–organic nitrogen ratios of sediments in a polluted estuary. J. Water Pollution Contr. Fed. 41:2, 1969.
Flentje, M. E. Instrumented techniques for water analysis. Water Works and Wastes Eng. 2(6):128, 1965.
Frolander, H. F. Biological and chemical features of tidal estuaries. J. Water Pollution Contr. Fed. 36:1037, 1964.
Galetti, B. J., and F. C. Snowden. New analysis instruments aid pollution control. Environ. Sci. Technol. 3(1):34, 1969.
Gjessing, E., and G. F. Lee. Fractionation of organic matter in natural waters on sephadex columns. Environ. Sci. Technol. 1(8):631, 1967.
Goodnight, C. J., and L. S. Whitley. Oligochaetes as indicators of pollution. Proc. 13th Ind. Waste Conf., Purdue Univ., 139, 1960.
Graham, E. D., P. G. Stoddart, and T. W. Severn. Plutonium monitoring techniques for 2 PR-111. Surface Contamination Proc. Inst. Symposium, Gatlinburg, Tenn., 2938, 1964.
Grasshoff, K. On the determination of silica in sea water. Deep Sea Res. II, 597, 1964.
Gunnerson, C. G. Optimizing sampling intervals in tidal estuaries. Trans. Amer. Soc. Civil Eng.–J. Sanitary Eng. Div., 92:103, 1966.
Hanes, N. B., and R. S. Tragala. Effect of laboratory storage of sea water on coliform. Water and Sewage Works 115(6):254, 1968.
Henley, D. E., W. H. Glaze, and J. K. G. Silvey. Isolation and identification of an odor compound produced by a selected aquatic actinomycete. Environ. Sci. Technol. 3(3):268, 1969.
Henriksen, A. An automatic method for determining nitrate and nitrite in fresh and saline waters. Analyst (London) 90:88, 1965.
Holly, E. R., and D. F. R. Harleman. Dispersion of pollutants in estuary-type flows. Mass. Inst. Technol., Dept. Civil Eng., Hydrodynamics Lab, Report No. 74, 202, 1965.
Holm-Hansen, O., W. H. Sutcliffe, Jr., and J. Sharp. Measurement of deoxyribonucleic acid in the ocean and its ecological significance. Limnol. Oceanog. 13:507–514, 1968.

Jenkins, D., and L. L. Medsker. Brucine method for determination of nitrate in ocean, estuarine, and fresh waters. Anal. Chem. 36:610, 1964.

Johnson, C. R., P. H. McClelland, and R. L. Boster. Rapid volumetric determination of sulphide in estuarine and sea waters. Anal. Chem. 36:300, 1964.

Joyner, T., M. L. Healy, D. Chakravarti, and T. Koyanagi. Preconcentration for trace analysis of sea waters. Environ. Sci. Technol. 1(5):417, 1967.

Kahn, L., and T. T. Brezenski. Determination of nitrate in estuarine waters—automatic determination using a brucine method. Environ. Sci. Technol. 1(6):492, 1967.

Kahn, L., T. T. Brezenski, and T. Frances. Determination of nitrate in estuarine waters. Comparison of a hydrazine reduction and a brucine procedure. Environ. Sci. Technol. 1(6):488, 1967.

Kammerer, P. A., and G. T. Lee. Freeze concentration of organic compounds in dilute aqueous solutions. Environ. Sci. Technol. 3(3):276, 1969.

Kawahara, T. K. Identification and differentiation of heavy residual oil and asphalt pollutants in surface waters by comparative ratios of infrared intensities. Environ. Sci. Technol. 3(2):150, 1969.

Konnov, V. A. Determination of nitrates and ammonia in sea water. Trudy Inst. Okeanol. 54:123, 1962.

Koyama, T., and T. G. Thompson. Identification and determination of organic acids in sea water by partition chromatography. J. Oceanogr. Soc. Japan 20:209, 1964.

Kreg, J., and K. H. Szekielda. Determination of organic carbon in sea water by using a new apparatus for the determination of very small amounts of carbon dioxide. Z. Anal. Chem. 207:330, 1965.

Leeds, J. V. Accuracy of discrete models used to predict estuary pollution. Water Resources Res. 3(2):481–490, 1967.

Ludwig, H. F., and B. Onodera. Collation, evaluation, and presentation of scientific and technical data relative to the marine disposal of liquid wastes. Report to California Water Quality Control Board, 105 pp. and Appendices, Sacramento, Calif., 1964.

Ludwig, H. F., and B. Onodera. Scientific parameters of marine waste discharge. Vol. III, Proc. First International Conference on Water Pollution Research, Pergamon Press, pp. 37-49, 1964.

Marcie, T. G. X-ray fluorescent analysis of trace toxic elements in water. Environ. Sci. Technol. 1(2):164, 1967.

Marten, J. F. The development of a multiple simultaneous actoanalyzer system for remote monitoring of water quality parameter. Ind. Chem. Belg. 32(12):1319, 1967.

Martin, D. F. Analysis of polluted waters. In Marine Chemistry 1. Marcel Dekker, Inc., New York, 1968.

Miyake, Y., I. Sugiura, and T. Nannite. Studies of radioactive contamination in local marine resources. Inst. Atom. Energ. Agency IAE-R-19, 138, 1962.

Montgomery, H. A. C. and C. Quarmby. The extraction of gases dissolved in water for analysis by gas chromatograph. Lab. Pract. 15:538, 1966.

Morris, A. W., and J. P. Riley. The determination of nitrate in sea water. Anal. Chem. Acta 29:272, 1963.

Murray, B. Desalting sea water with ammonia, Part I: Ion Exchange. Water and Sewage Works 115(10):482, 1968.

Murray, B. Desalting sea water with ammonia, Part II: Osmosis. Water and Sewage Works 115(11):525, 1968.

MacIsaac, J. J. Ammonia determinations by two methods in the northeast equatorial Pacific Ocean. Limnol. Oceanog. 12:552–554, 1967.

O'Connor, D. J. Estuarine distribution of non-conservative substances (mathematical treatment, pollution). Amer. Soc. Civil Eng., J. Sanit. Eng. Div. 93(SA4):115, 1967.

Oglesby, R. T., and R. O. Sylvester. Marine biological monitoring of oil refinery liquid waste emissions. Proc. 19th Ind. Waste Conf., Purdue Univ., 1964.

Orlando, G., and V. Riva. Determination of strontium-90 and caesium-137 in sea water sampled simultaneously from several bodies of water facing the province of Genoa. Ig. Med. Prevent. 6:73–77, 1965.

Palmork, K. H. Studies of the dissolved organic compounds in the sea. Biol. Abstr. 45:7462, 1964.

Pamalmat, M. M., and K. Banse. Oxygen consumption by the seabed II. *In situ* measurements to a depth of 180 m. Limnol. Oceanog. 14:250–259, 1969.

Patrick, R. The structure of diatom communities under varying ecological conditions. Annals of the New York Academy of Sciences 108(2):359, 1963.

Patten, R. Species diversity in net phytoplankton of Raritan Bay. J. Mar. Res. 20:57, 1962.

Pearson, E. A., P. N. Storrs, and R. E. Selleck. Waste Discharges and Loadings, Vol. III, Final Report Series, Comprehensive Investigation of San Francisco Bay. University of California Sanitary Engineering Research Laboratory Publication, June 1969.

Pomeroy, L. R., E. E. Smith, and C. M. Grant. The exchange of phosphate between estuarine water and sediments. Limnol. Oceanog. 10:167–172, 1965.

Preview of the Houston Pollution Control Exposition Exhibits. Environ. Sci. Technol. 3(4):343, 1969.

Rambow, Carl A. Pollution study of a future tidal estuary. J. Water Pollution Contr. Fed. 36(4):520, 1964.

Raymont, J. E. G., and J. Shields. Toxicity of copper and chromium in the marine environment. Int. J. Air Water Pollution 1:435–443, 1963.

Riley, J. P., and M. Tonguda. The lithium content of sea water. Deep Sea Res. 11:363–368, 1964.

Rosen, A. A., and M. Rubin. Discriminating between natural and industrial pollution through carbon dating. J. Water Pollution Contr. Fed. 37(9):1302–1307, 1965.

Ryther, J. H. The ecology of the phytoplankton blooms in Moriches Bay and Great South Bay, Long Island, N.Y. Biol. Bull. Mar. Biol. Lab., Woods Hole, 106: 198–209, 1954.

Sabo, J. J., and P. H. Bedrosian. Studies of the fate of certain radionuclides in estuarine and other aquatic environments. U.S. Publ. Hlth. Serv. Publ. No. 999-R-3, 1963.

Selleck, R. E. Persistence of spent sulphite waste liquor in estuarine environment. Develop. Ind. Microbiol. 5:71–77, 1963.

Selleck, R. E., and E. A. Pearson. The San Francisco Bay water pollution investigation. Prepr. 4th Ind. Waste Conf., Tex. Water Pollut. Contr. Ass., B-20-31, 1964.

Seki, H., J. Skelding, and T. R. Parsons. Observations on the decomposition of a marine sediment. Limnol. Oceanog. 13:440–447, 1968.

Seymour, A. H., ed. Radioactivity in the marine environment. National Academy of Sciences–National Research Council, Washington, D.C., (in preparation).

Shaffer, L. H., and R. A. Knight. Nucleation of chrystalline phases from sea water concentrates. Environ. Sci. Technol. 1(8):661, 1967.

Sieburth, J. McN., and J. T. Conover. Slicks associated with trichodesmium blooms in the Sargasso Sea. Nature 205:830–831, 1965.

Skrinde, R. T., and H. D. Tomlinson. Organic micropollution instrumentation and measurement. J. Water Pollution Contr. Fed. 35(10):1292–1306, 1963.

Smith, R. C., K. C. Pellai, T. J. Chow, and T. R. Folsom. Determination of rubidium in sea water. Limnol. Oceanog. 19:226–232, 1965.

Smith, R. H. Survey of organic content in bottom sediments of the Houston Ship Channel. M.S. thesis, Texas A&M Univ., 1968.

Snowden, F. C. Instrumentation for pollution monitoring and control. Proc. of the Texas A&M 23rd Annual Symposium on Instrumentation for the Process Industries, 1968.

Spena, A. Disposal of radioactive wastes in the sea. Ann. Med. Nav. Trop. 67:463–483, 1962.

Strandberg, C. H. Water quality analysis. Photogrammetric Eng. 32(2):234–248, 1966.

Strickland, J. D. H. A manual of sea water analysis. Bull. Fish. Res. Board Can. 125, 1965.

Szabo, B. J., and O. Joensuu. Emission spectrographic determination of barium in sea water using a cation exchange concentration. Environ. Sci. Technol. 1(6):499, 1963.

Szekielda, K. H., and J. Krey. The determination of particulate, organically-bound carbon in sea water by a new rapid method. Mikrochim. Acta 149–159, 1965.

Torpey, W. N. Effects of reducing pollution of the Thames. Water and Sewage 115(7):295, 1968.

Templeton, W. L. Ecological aspects of the disposal of radioactive wastes to the sea. In Ecology and the Industrial Society, A Symposium of the British Ecological Society, Swansea, p. 404, 1965.

Vaccaro, R. F., and H. W. Jannasch. Variations in uptake kinetics for glucose by natural populations in seawater. Limnol. Oceanog. 12:540–541, 1967.

Willingham, C. A., and K. J. Anderson. Use of microorganisms for detecting toxic materials in water. Water and Sewage Works 114(1):25, 1967.

Yen, L. Determination of nitrate in a sea water with sodium diphenylbenzidine sulfonate. Oceanologia Limnol. 5:115–123, 1963.

Van Lopik, J. R., G. S. Rambie, and A. F. Pressman. Pollution surveillance by noncontact infrared techniques. J. Water Pollution Contr. Fed. 40(1):425–438, 1968.

Volkov, V. G. Instrument for measurement of the distribution of salinity and temperature of sea water. Trudy Inst. Okeanol. 67:216–229, 1964.

Wangersky, P. J. Organic chemistry of sea water. Amer. Scientist 53(3):358, 1965.

Warnick, S. L., and A. R. Gaufin. Determination of pesticides by electron capture gas chromatography. Amer. Water Works Ass. 57(8):1023, 1965.

Wass, M. L. Indicators of pollution. In Pollution and Marine Ecology. Interscience Publishers, New York, 1967.

Wastler, T. A. Measuring estuarine pollution. Oceanology 43, May/June 1969.
Wilkinson, L. Nitrogen transformation in a polluted estuary. Water Pollution 7:737, 1963.

PHYSICAL PROCESSES AND INTERACTIONS

Abraham, G. Jet diffusion in liquid of greater density. Proc., J. Hydraul. Div., A.S.C.E., 86, No. HY6. Proc. paper 2506, 1–13, 1960.

Abraham, G. Horizontal jets in stagnant fluid of other density. Proc., J. Hydraul. Div., A.S.C.E., 91, No. HY4. Proc. Paper 4411, 139–154, 1965.

Allan Hancock Foundation. An investigation on the fate of organic and inorganic wastes discharged into the marine environment and their effects on biological productivity. State of California Water Quality Control Board Publ. No. 29, 117 pp., 1965.

Batchelor, G. K. Diffusion in a field of homogeneous turbulence. II. The relative motion of particles. Proc. Comb. Phil. Soc. 48:345–362, 1952.

Bowden, K. F. The mixing processes in a tidal estuary. Int. J. Air Water Pollution 7(4–5):343–356, 1963.

Bowden, K. F. Horizontal mixing in the sea due to a shearing current. J. Fluid Mech. 21(1):83–95, 1965.

Brezenski, F. T., R. Russomanno, and P. DeFalco, Jr. The occurrence of *Salmonella* and *Shigella* in post-chlorinated and non-chlorinated sewage effluents and receiving waters. Health Lab. Sci. 2:40–46, 1965.

Brooks, N. H. Diffusion of sewage effluent in an ocean current. Proc. First Int. Conf. on Waste Disposal in the Marine Environment, Pergamon Press, New York, 1959.

Buelow, Ralph W. Ocean disposal of waste material. Trans. Nat. Symposium on Ocean Sciences and Engineering of the Atlantic Shelf, Philadelphia, March 19–20, 1968, Marine Technology Society, pp. 311–337, 1968.

Bumpus, Dean F. Circulation on the continental shelf of the east coast of the U.S. 5th Cont. Shelf Workshop, USNA, Annapolis, Md., December 16–17, 1968.

Cameron, W. M., and D. W. Pritchard. Estuaries. In The Sea, Vol. 2, M. N. Hill, ed. Interscience, New York, pp. 306–324, 1963.

Carpenter, J. H. The Chesapeake Bay Institute study of the Baltimore Harbor. Proc. 33rd Ann. Conf. Maryland–Delaware Water and Sewage Ass., 62–78, 1960.

Carter, H. H. A method for delineating an exclusion area around a sewage outfall in a tidal estuary based on water quality with application to the Severn and Choptank rivers. Special Report No. 9, Chesapeake Bay Institute, The Johns Hopkins University, Reference 65-3, October 1965.

Carter, H. H., and A. Okubo. A study of the physical processes of movement and dispersion in the Cape Kennedy area. Final Report under the U.S. Atomic Energy Commission Contract, Rept. No. NYO-2973-1, Chesapeake Bay Institute, The Johns Hopkins University, 1965.

Carter, H. H., D. W. Pritchard, and J. H. Carpenter. The design and location of a diffuser outfall for a municipal waste discharge at Ocean City, Maryland. Chesapeake Bay Institute Special Report No. 10, Ref. 66-2, 1966.

Carter, R., and W. Kim. Determination and removal of floatable material from wastewater. Engineering-Science, Inc., U.S. Public Health Service Contract 120-64, Nov. 1965.

Chick, H. Investigation of the laws of disinfection. J. Hyg., 8:92, 1908.

Costin, M., P. Davis, R. Gerard, and B. Katz. Dye diffusion experiments in the New York Bight. Tech. Rept. No. Cu-2-63, Lamont Geological Observatory, Columbia Univ., 18 pp. (unpublished), 1963.

Csanady, G. T. Accelerated diffusion in the skewed shear flow of lake currents. J. Geophys. Res. 71(2):411–420, 1966.

Fan, Loh-Nien. Turbulent buoyant jets into stratified or flowing ambient fluids. W. M. Keck Laboratory of Hydraulics and Water Resources, California Institute of Technology, Report No. KH-R-15, June 1967.

Fischer, H. B. The mechanics of dispersion in natural streams. J. Hydraul. Div., Proc. A.S.C.E. 93 (HY 6):187–216, 1967.

Fischer, H. B. Methods for predicting dispersion coefficients in natural streams, with applications to lower reaches of the Green and Duwamish rivers, Washington. Geological Survey Professional Paper 582-A, U.S. Government Printing Office, Washington, D.C., 1968.

Foxworthy, J. E., R. B. Tibby, and G. M. Barsom. Dispersion of a waste field in the sea. J. Water Pollution Contr. Fed. 1170–1193. July 1966.

Foxworthy, J. E. Eddy diffusion and the four-thirds law in near-shore coastal waters. Univ. of Southern California, Allan Hancock Foundation, Report 68-1, 72 pp., 1968.

Foxworthy, J. E. Turbulent diffusion and bacterial reduction in waste fields in the ocean. Univ. of Southern California, Allan Hancock Foundation, Report 68-2, 1968.

Frankel, R. J., and Cumming, J. D. Turbulent mixing phenomenon of outfalls. Proc., J. Sanit. Eng. Div., A.S.C.E., 91, No. SA2. Proc. paper 4297, 33–59, 1965.

Hansen, D. V., and M. Rattray, Jr. Gravitational circulation in straits and estuaries. J. Marine Res. 23:104–122, 1965.

Harleman, D. R. F. The significance of longitudinal dispersion in the analysis of pollution in estuaries. Proc. 2nd Int. Water Pollution Research Conference, Tokyo. Pergamon Press, New York. 1965.

Holly, E. R. Unified view of diffusion and dispersion. J. Hydraul. Div., A.S.C.E., 95 (HY 2) March 1969.

Harremoes, P. Diffuser design for discharge to a stratified water. 4th Int. Conf. on Water Pollution Research. Pergamon Press, Ltd., New York, 1968.

Hyperion Engineers, Inc. Ocean outfall design. Rept. to City of Los Angeles, 1957. See review in J. Sanit. Eng. Div., A.S.C.E., 87 (SA 4):1–32, July 1961.

Ippen, A. T., and D. R. F. Harleman. One-dimensional analysis of salinity intrusion in estuaries. Tech. Bull. No. 5, Comm. Tidal Hydraulics. U.S. Army Corps of Engineers, 52 pp., 1961.

Ippen, A. T., and G. B. Keulegan. Salinity intrusion in estuaries. In Evaluation of present state of knowledge of factors affecting tidal hydraulics and related phenomena. Report No. 3, Comm. Tidal Hydraulics. U.S. Army Corps of Engineers, 1965.

Iselin, C. O. D. Coastal currents and the fisheries. Papers in Marine Biology and Oceanography. Suppl. to Vol. 3 of Deep-Sea Research, pp. 474–478, 1955.

Joseph, J., and H. Sender. Uber die Horizontale Diffusion Im Meere. Dt. Hydrogr. 2, 11:49–77, 1958.

Ketchum, B. H., and D. Jean Keen. The accumulation of river water over the continental shelf between Cape Cod and Chesapeake Bay. Papers in Marine Biology and Oceanography. Suppl. to Vol. 3 of Deep-Sea Research, pp. 346–347, 1955.

Kullenberg, G. *In situ* measurements of horizontal and vertical diffusion in the thermocline in Swedish coastal waters. Symp. on Diffusion, Int. Ass. Phys. Oceanog., IUGG, Sept. 25–Oct. 7, 1967, held in Berne, Switzerland, 1967.

Leendertze, J. J. Computational inputs of a computational model for long water wave propagation. The Rand Corporation, RM 5294-PR, May 1967.

Ludwig, H. F., and R. Carter. Analytical characteristics of oil-tar material on Southern California beaches. J. Water Pollution Contr. Fed. 33:1123, 1961.

Okubo, A. A review of theoretical models of turbulent diffusion in the sea. J. Oceanogr. Soc. Japan, 20th Anniv. Vol., 286–320, 1962.

Okubo, A. Horizontal diffusion from an instantaneous point-source in the sea. Chesapeake Bay Institute Technical Report 32, 123 pp., 1962.

Okubo, A. A new set of diffusion diagrams. Chesapeake Bay Institute Technical Report 38, 35 pp., 1968.

Okubo, A. Some remarks on the importance of the "shear effect" on horizontal diffusion. J. Oceanogr. Soc. Japan 24:60–69, 1968.

Okubo, A., and M. J. Karweit. Diffusion from a continuous source in a uniform shear flow. Limnol. and Oceanog. 14:514–520, 1969.

Ozmidov, R. V. On the calculation of horizontal turbulent diffusion of the pollutant patches in the sea. Doklady Akad. Nauk. SSSR 120:761–763, 1958.

Pearson, E. A. An investigation of the efficacy of submarine outfall disposal of sewage and sludge. Publ. No. 14, Calif. Water Pollution Control Board, 154 pp., 1955.

Pearson, E. A., and R. Carter. On eddy diffusivity and current determination in outfall design. Presented at Manhattan College 9th Summer Institute, New York, 1964.

Phelps, E. P. Stream sanitation. John Wiley and Sons, New York, 211 pp., 1944.

Pickard, G. L. Oceanographic features of inlets in the British Columbia mainland coast. J. Fish. Res. Board Can. 18:907–999, 1961.

Price, W. A., and M. P. Kendrick. Field and model investigation into the reasons for siltation in the Mersey Estuary. Proc. Inst. Civil Engrs. (London) 24:473–518, 1963.

Pritchard, D. W. The physical structure, circulation, and mixing in a coastal plain estuary. Chesapeake Bay Institute Technical Report 3, 55 pp., 1952.

Pritchard, D. W. Estuarine circulation patterns. Proc. A.S.C.E., Hydraulic Division 81:717, 1955.

Pritchard, D. W. Problems related to disposal of radioactive wastes in estuarine and coastal waters. Trans. 2nd Seminar on Biological Problems in Water Pollution, held April 20–24, 1959.

Pritchard, D. W. The application of existing knowledge to the problem of radioactive waste disposal into the sea. Disposal of Radioactive Wastes 2, Int. Atomic Energy Agency, Vienna, pp. 229–248, 1960.

Pritchard, D. W., and J. H. Carpenter. Measurements of turbulent diffusion in estuarine and inshore waters. Bull. Int. Ass. Sci. Hydrol. No. 20, pp. 37–50, 1960.

Pritchard, D. W., A. Okubo, and H. H. Carter. Observations and theory of eddy movement and diffusion of an introduced tracer material in the surface layers of the sea. Disposal of Radioactive Wastes into Seas, Oceans, and Surface Waters. Int. Atomic Energy Agency, Vienna, pp. 397–424, 1966.

Pritchard, D. W. What is an estuary: Physical viewpoint. In Estuaries, G. H. Lauff (ed.), Amer. Ass. Advan. Sci., Washington, D.C., pp. 3–5, 1967.

Pritchard, D. W. Dispersion and flushing of pollutants in estuaries. J. Hydrol. Div. Proc. Amer. Soc. Civil Eng. 95 (HY 1):115–124, 1969.

Proudman, J. Dynamical oceanography. Wiley, New York, 409 pp., 1953.

Rattray, M., Jr. Some features of the dynamics of the circulation in fjords. Proc. Conf. on Estuaries, Jekyll Island, 1964.

Rawn, A. M., F. R. Bowerman, and N. H. Brooks. Diffusers for disposal of sewage in sea water. Proc., J. Sanit. Eng. Div., A.S.C.E., 86 (SA 2):65–105, 1960.

Saelen, O. H. Some features of the hydrography of Norwegian fjords. Proc. Conf. on Estuaries, Jekyll Island, 1964.

Schultz, E. A., and H. B. Simmons. Fresh water–salt water density currents, a major cause of siltation in estuaries. Tech. Bull. No. 2, Comm. Tidal Hydraulics. U.S. Army Corps of Engineers, 28 pp., 1957.

Spino, D. F. Elevated temperature technique for isolation of *Salmonella* from streams. Appl. Microbiol. 14:591–596, 1966.

Stommel, H. Computation of pollution in a vertically mixed estuary. Sewage Ind. Wastes 25:1065–1071, 1953.

Stommel, H., and H. G. Farmer. Abrupt change in width in two-layer open channel flow. J. Marine Res. 11:205–214, 1952.

Taylor, G. I. Dispersion of soluble matter in solvent flowing slowly through a tube. Proc. Roy. Soc. London, A 219, 186–203, 1953.

Taylor, G. I. The dispersion of matter in turbulent flow through a pipe. Proc. Roy. Soc. London, A 223:446–468, 1954.

CHEMICAL FACTORS

Adams, D. D., and F. A. Richards. Dissolved organic matter in an anoxic fjord, with special reference to the presence of mercaptans. Deep-Sea Res. 15:471–481, 1968.

Allan Hancock Foundation. An investigation on the fate of organic and inorganic wastes discharged into the marine environment and their effects on biological productivity. State of California Water Quality Control Board Publ. No. 29, 117 pp., 1965.

Allan Hancock Foundation. An oceanographic and biological survey of the Southern California mainland shelf. State of California Water Quality Control Board Publ. No. 27, 232 pp., 1965.

American Chemical Society. Equilibrium concepts in natural water system. Advances in Science Series, No. 67, 1967.

American Chemical Society. Trace inorganics in water. Advances in Science Series, No. 73, 1968.
AWWA Task Group. Sources of nitrogen and phosphorus in water supplies. J. Amer. Water Works Ass., p. 344, March 1967.
Barrett, M. J., D. Munro, and A. R. Agg. Radiotracer dispersion studies in the vicinity of a sea outfall. Paper III-13, 4th Int. Conf. on Water Pollution Research, Prague, 1969.
Berner, R. A. An idealized model of dissolved sulfate distribution in recent sediments. Geochim. Cosmochim. Acta, 28:1497–1503, 1964.
Bien, G. S., D. E. Contois, and W. H. Thomas. Removal of soluble silica from fresh water entering the sea. Geochim. Cosmochim. Acta, 14:35–45, 1958.
Blumer, M. Dissolved organic compounds in sea water. Saturated and olifinic hydrocarbons, singly branched fatty acids. Symposium on Organic Matter in Natural Waters, College, Alaska, Sept. 2-4, 1968.
Bowen, H. J. M. Trace elements in biochemistry. Academic Press, London and New York, 1966.
Burrell, D. C. and D. W. Hood, eds. Report No. SAN-310 3-13 to the U.S. Atomic Energy Commission, Institute of Marine Science, Univ. of Alaska, Part II, pp. 108–115, 1969.
Callame, G. Sur la discussion des gaz a l'interieur des sediments marins. Compt. Rend. 260:1220–1225, 1965.
Carritt, D. E., and S. Goodgal. Sorption reactions and some ecological implications. Deep-Sea Res. 1:224–248, 1954.
Carroll, D. Ion exchange in clays and other minerals. Bull. Geol. Soc. Amer. 70: 749–80, 1959.
Chave, K. E. Carbonate-organic interactions in seawater. Symposium on Organic Matter in Natural Waters, College, Alaska, Sept. 2-4, 1968.
Chester, R. Adsorption of zinc and cobalt on illite in sea water. Nature 206:884–6, 1965.
Cruickshank, M. J., C. M. Ramanowitz, and M. P. Overall. Offshore mining—present and future. Eng. and Min. J. 169:84–91, 1968.
Degens, E. T. Molecular nature of nitrogenous compounds in sea water and recent marine sediments. Symposium on Organic Matter in Natural Waters, College, Alaska, Sept. 2-4, 1968.
Duke, T. W., J. N. Willis, and D. A. Wolfe. A technique for studying the exchange of trace elements between estuarine sediments and water. Limnol. Oceanogr. 13:541–545, 1968.
Eicholz, G. G., T. F. Craft, and Ann N. Galli. Trace element fractionation by suspended matter in water. Geochim. Cosmochim. Acta 31:737, 1967.
Feth, J. H., G. E. Roberson, and W. L. Polzer. U.S. Geol. Survey Water Supply Paper No. 1535, 1964.
Garrels, R. M., and M. E. Thompson. A chemical model for sea water at $25°C$ and one atmosphere total pressure. Amer. J. Sci. 260:57–66, 1962.
Garrett, W. D. The organic chemical composition of the ocean surface. Deep-Sea Res. 14:221, 1967.
Haven, D. S., and R. Morales-Alamo. Use of fluorescent particles to trace oyster biodeposits in marine sediments. J. Conseil Perm. Intern. Explor. Mer. 30:267–269, 1966.

Hood, D. W., ed. Impingement of man on the ocean. J. Wiley-Interscience, New York, (in press).
Kanwisher, J. On the exchange of gases between the atmosphere and the sea. Deep-Sea Res. 10:195-207, 1963.
Krauskopf, K. B. Factors controlling the concentrations of thirteen rare metals in seawater. Geochim. Cosmochim. Acta 9:1-32, 1956.
Lauff, G. H., ed. Estuaries. Publ. No. 83, Amer. Ass. Advan. Sci., Washington, D.C., 1967.
Lowman, F. G., D. M. Phelps, R. McClin, F. Roman de Vega, I. O. de Padovani, and J. Garcia. Interactions of the environmental and biological factors on the distribution of trace elements in the marine environment. Proc. Symposium. Disposal of Radioactive Wastes into Seas, Oceans, and Surface Waters, Vienna, May 16-20, 1966.
MacKenzie, F. T., and R. M. Garrels. Chemical mass balance between rivers and oceans. Amer. J. Sci. 264:507, 1966.
Manheim, F. T. A geochemical profile in the Baltic Sea. Geochim. Cosmochim. Acta 25:52-70, 1961.
Menzel, D. L. Particulate organic carbon in the deep sea. Deep-Sea Res. 14:229, 1967.
Natarajan, K. V., and R. C. Dugdale. Bioassay and distribution of thiamine in the sea. Limnol. Oceanog. 11:621, 1966.
Olson, T. A. and F. A. Burgess, eds. Pollution and marine ecology. Interscience, New York, 1967.
Onufrienok, I. P., and R. S. Solodovnikova. The influence of humus on the behavior of the microcomponents in natural waters. Trudy Romsk. Gos. Univ., Ser. Khim. (USSR), 170, 163, 1964; Chem. Abstr. 63:1581, 1965.
Reesburgh, W. S. Observations of gases in Chesapeake Bay sediments. Limnol. Oceanogr. 14:368-375, 1969.
Richards, F. A. Chemical oceanography. Trans. Amer. Geophys. Union, 48:595-604, 1967.
Riley, J. P., and J. Skirrow, eds. Chemical oceanography. Academic Press, London, Vol. 1, pp. 611-646, 1965.
Rudd, R. L. Pesticides and the living landscape. Univ. of Wisconsin Press, Madison, 1964.
Sanders, H. L., P. C. Mangelsdorf, Jr., and G. R. Hampson. Salinity and faunal distribution in the Pocasset River, Massachusetts. Limnol. Oceanogr. Redfield Anniversary Volume, P 216-R 229, 1965.
Schubel, J. R. Suspended sediment of the northern Chesapeake Bay. Tech. Report No. 35, Chesapeake Bay Institute, The Johns Hopkins Univ., 263 pp., 1968.
Sears, M., ed. Oceanography. AAAS, Washington, D.C., 1961.
Sillen, L. G. The ocean as a chemical system. Science 156:1189, 1967.
Slowey, J. F., Lela M. Jeffrey, and D. W. Hood. Evidence for organic complexed copper in sea water. Nature 214:377-378, 1967.
Spence, R. Extraction of uranium of sea water. Nature 1110-1115, Sept. 12, 1969.
Stefansson, U., and F. A. Richards. Processes contributing to the nutrient distributions off the Columbia River and Strait of Juan de Fuca. Limnol. Oceanog. 8:394-410, 1963.

Strickland, J. D. H., and T. R. Parsons. A practical handbook of seawater analysis. Bulletin 167. Fisheries Research Board of Canada, Ottawa, 1968.
Werner, A. E., and M. Waldichuk. A sampler for gases in bottom sediments. Limnol. Oceanog. 158–161, 1967.

BIOLOGICAL EFFECTS

Allan Hancock Foundation. An investigation on the fate of organic and inorganic wastes discharged into the marine environment and their effects on biological productivity. State of California Water Quality Control Board Publ. No. 29, 117 pp., 1965.
Allan Hancock Foundation. An oceanographic and biological survey of the Southern California mainland shelf. State of California Water Quality Control Board Publication No. 27, 232 pp., 1965.
Allen, G. H. An oceanographic study between the points of Trinidad Head and the Eel River. State of California Water Quality Control Board Publication No. 25, 135 pp., 1964.
Armstrong, N. E., E. F. Gloyna, and B. J. Copeland. Ecological aspects of stream pollution. In Advances in water quality improvement, E. F. Gloyna and W. W. Eckenfelder, Jr., eds. Univ. of Texas Press, Austin, 1968.
Bartsch, A. F., and W. M. Ingram. Stream life and the pollution environment. Public Works Publications, Ridgewood, N.J., 1959.
Burd, R. S. A study of sludge handling and disposal. FWPCA Publ. WP-20-4, Water Pollution Control Research Series, 326 pp., 1968.
Cooper, E. L., ed. A symposium on water quality criteria to protect aquatic life. Trans. Amer. Fish. Soc. Spec. Publ. No. 4, 1967.
Glossary Water and Wastewater Control Engineering. ASCE, AWWA, 1969.
Hynds, H. B. N. The Biology of Polluted Waters. Liverpool University Press, Liverpool, 202 pp., 1963.
Kinn, O., ed. International symposium 1967 on biological and hydrographical problems of water pollution in the North Sea and adjacent waters. In Meeresuntersuchungen, Vol. 17, 1968.
Lackey, J. B. Nutrient and pollutant response of estuarine biotas. Tech. Paper No. 408. Proc. National Symposium on Estuarine Pollution, pp. 1–11, 1968.
Ludwig, H. F., and B. Onodera. Scientific parameters of marine waste discharge. Presented at International Conf. on Water Pollution Research, London, Sept. 3–7, 1962.
Ludwig, H. F., E. Kazmierczak, and R. Carter. Waste disposal and the future at Lake Tahoe. Proceedings of the ASCE, Paper No. 3947, June 1964.
Ludwig, H. F., W. D. Bishop, and R. Carter. 1964. A study of beach pollution in tidal estuaries—a discussion. Presented at 2nd International Water Pollution Research Conf., Tokyo, August 1964.
Ludwig, H. F., R. C. Carter, and J. Scherfig. Characterization of oils in the sea. Symp. Poll. mar. Micro-org. Prod. petrol., Monaco, 1964.
Ludwig, H. F., and J. Scherfig. Determination of floatables and hexane extractables in sewage. 3rd International Conf. on Water Pollution Research, Munich, Sept. 1967.

Ludwig, H. F., and R. F. Smith. Eutrophication mechanisms at Lake Tahoe. In Water research, Pergamon Press, Vol. 2, 1968.

Ludwig, H. F., P. N. Storrs, E. A. Pearson, R. Walsh, and E. J. Stann. Estuarine water quality and biologic population indices. Presented at 4th International Conf. on Water Pollution Research, Prague, April 1969.

Ludwig, H. F. On the survey and prediction of pollution at the Omuta Industrial Harbor. Discussion presented at 4th International Conf. on Water Pollution Research, Prague, April 1969.

Ludwig, H. F. Effects of waste disposal into marine waters. Univ. of Calif. Water Resources Engineering Education Series, San Francisco, January 28–30, 1970.

McLeese, D. W. Effects of temperature, salinity and oxygen on the survival of the American lobster. J. Fish. Res. Bd. Can. 13:247–272, 1956.

Mackenthun, K. M., W. M. Ingram, and R. Porges. Limnological Aspects of Recreational Lakes. Public Health Service Publ. No. 1167, 1964.

Mann, H. Effects on the flavour of fishes by oils and phenols. Symp. Poll. mar. Micro-org. Prod. petrol, Monaco 1964, 371–374, 1965.

Nelson-Smith, A. The effects of pollution and emulsifier cleansing on shore life in south-west Britain. J. Appl. Ecol. 5:97–107, 1968.

Nelson-Smith, A. A classified bibliography of oil pollution: The biological effects of oil pollution on littoral communities. Supplement to Field Studies, Vol. 2, Field Studies Council, pp. 165–195, 1968.

North, W. J., M. Neushul, and K. A. Clendenning. Successive biological changes in a marine cove exposed to a large spillage of mineral oil. Symp. Poll. mar. Microorg. Prod. petrol., Monaco 1964, 335–354. 1965.

Odum, E. P., and H. T. Odum. Fundamentals of ecology. W. B. Saunders Co., Philadelphia, 1959.

Odum, H. T., and C. M. Hoskin. Comparative studies on the metabolism of marine waters. Publ. Inst. Mar. Sci. 5:16–46, 1958.

Odum, H. T. Analysis of diurnal oxygen curves for the assay of reaeration rates and metabolism in polluted marine bays. In Waste disposal in the marine environment, E. Pearson (ed.). Pergamon Press, New York, 1960.

Odum, H. T., J. E. Cantlon, and L. S. Kornicker. An organization hierarchy postulate for the interpretation of species–individual distributions, species entropy, ecosystem evolution, and the meaning of a species–variety index. Ecology 41:395–399, 1960.

Odum, E. P. The strategy of ecosystem development. Science 164:262–270, 1969.

Orlob, G. T. Effects of digested sludge discharge on the ocean environment near the City of San Diego outfall. Report of an investigation for San Diego Regional Water Pollution Control Board. Water Resources Engineers, Inc., 1965.

Pearson, E. A., ed. Proceedings, 1st International Conference on Waste Disposal in the Marine Environment. Pergamon Press, New York, 1959.

Pearson, E. A. Notes on bioassay methodology. Water Pollution Panel, Pacific Fisheries Biologist Conference, Richardson Springs, Calif., March 8, 1966.

Pearson, E. A. The case for continuous flow (chemostats) kinetic descriptions of plankton-nutrient growth relationships in eutrophication analyses. Prepared for Joint Industry Government Committee Meeting on Algae Growth Potential, Chicago, March 6–8, 1968.

Phelps, E. B. Stream sanitation. John Wiley & Sons, Inc., New York, 1944.

Pomeroy, R. D., and G. T. Orlob. Problems of setting standards and of surveillance for water quality control. State of California Water Quality Control Board Publ. No. 36, 1967.

Portier, P. Mécanisme de la mort des oiseaux dont le plumage est impregné de mazout. Bull. Soc. Nat. Accl., 11, 1934.

Portier, P. and A. Raffy. 1934. Mécanisme de la mort des oiseaux dont le plumage est impregné de carbures d'hydrogéne. C. R. Acad. Sci. Paris, 198:851–853,1934.

Portmann, J. E., and P. M. Connor. The toxicity of several oil-spill removers to some species of fish and shellfish. Marine Biology 1(4):322–329, 1968.

Radcliffe, D. R., and T. A. Murphy. Biological effects of oil pollution—bibliography. F.W.P.C.A. Dept. Interior Program No. 15080 FHU 10/69. 1969.

Ross, R. D. Industrial Waste Disposal. Reinhold Book Corp., New York, 340 pp., 1968.

Silliman, R. P. Population models and test populations as research tools. Bioscience 19(6):524–528, 1969.

Simpson, A. C. The Torrey Canyon disaster and fisheries. Min. Agric. Fish. Food Lab. Leaflet 18, 1968.

Smith, J. E., ed. Torrey Canyon pollution and marine life. Published for the M.B.A. by Cambridge University Press, 1968.

Standard methods for the examination of water and wastewater. APHA, AWWA, WPCF, 12th ed., 1965.

Transactions National Symposium. Ocean sciences and engineering of the Atlantic Shelf. Philadelphia, March 1920.

Turner, C. H., E. E. Ebert, and R. R. Ginen. The marine environment offshore of Point Loma, San Diego, California. A Report to the San Diego Regional Water Quality Control Board No. 9 from the California Department of Fish and Game, 1966.

Turner, C. H., A. R. Strackan, and C. T. Mitchell. Survey of a marine environment subsequent to installation of a submarine outfall. A Report to the San Diego Regional Water Quality Control Board No. 9 from California Department of Fish and Game, 1968.

Wilhm, J. L. Comparison of some diversity indices applied to populations of benthic macroinvertebrates in a stream receiving organic wastes. J. Water Pollution Contr. Fed. 39:1673–1683, 1967.